PLANNED INVASION OF JAPAN, 1945:

THE SIBERIAN WEATHER ADVANTAGE

Fig. 1. Map of countries and cities mentioned in the text, route of flights to Siberia origin

gton, D.C. and Seattle, WA, and the Trans-Siberian Railroad in the USSR and Manchuria.

PLANNED INVASION OF JAPAN, 1945:

THE SIBERIAN WEATHER ADVANTAGE

H. S. Yoder, Jr.

LCDR, U. S. Naval Reserve, Retired
Director Emeritus, Geophysical Laboratory,
Carnegie Institution of Washington

AMERICAN PHILOSOPHICAL SOCIETY
INDEPENDENCE SQUARE ■ PHILADELPHIA
1997

Memoirs
of the
American Philosophical Society
Held at Philadelphia
for Promoting Useful Knowledge
Volume 223

Jacket illustrations: Front—US Task Force maneuvering off Japan, 17 August 1945. Photographed by the USS Shangri-La (CV-38). The carrier in the lower right is the USS *Wasp* (CV-38). Photo No. 80-G-278815, National Archives. Back—Waves of landing craft approaching Iwo Jima, 19 February 1945. Mt. Suribachi is in the center. Photo No. 80-G-415308, National Archives, Back flap—Hatten S. Yoder, Jr. 1945, from the author.

ISBN: 0-87169-223-6

US ISSN: 0065-9738

Library of Congress Cataloging-in-Publication Data
Yoder, H. S. (Hatten Schuyler), Jr. 1921-
    Planned invasion of Japan, 1945 : The Siberian weather
advantage / H. S. Yoder, Jr.
        p.   cm.—(Memoirs of the American Philosophical Society
Held at Philadelphia for Promoting Useful Knowledge, ISSN 0065-
9738 ; v. 223)
    Includes bibliographical references and index.
    ISBN 0-87169-223-6 (hardbound)
    1. Yoder, H. S. (Hatten Schuyler), Jr. 1921–.   2. Military
meteorology—Russia (Federation)—Siberia.   3. World War,
1939–1945—Personal narratives, American.   4. World War,
1939–1945—Campaigns—Japan.   5. Meteorologists—United
States—Biography.   6. United States.   Navy—Biography.   7. Fleet
Weather Central Khabarovsk (Russia)   I. Title.   II. Series: Memoirs
of the American Philosophical Society ; v. 223.
    Q11.P612   vol. 223
    [D810.M42]
    081 s—dc21
    [940.54'8673]   97-22539
                                                                           CIP

CONTENTS

FIGURES

"Our past is only a little less uncertain than our future, and, like the future, is always changing, always revealing and concealing."

Daniel J. Boorstin (1987, p. ix)
Historian, Librarian of Congress Emeritus
Member American Philosophical Society

FOREWORD

by

Commanding Officer of the MOKO Expedition to Siberia,
1945–1946

A NUMBER OF ARTICLES have been written about the Fleet Weather Central in eastern Russia during World War II. The following story is from a different viewpoint. Reading it shows you how difficult it is to accomplish your mission when your every movement is suspect, and when Soviet personnel with whom you have to collaborate cannot make a decision without the approval of their superiors.

I have high praise for all the personnel of the Fleet Weather Central for their loyalty, moral conduct, and ingenuity. In spite of all the trials and frustrations, I did not have a single case of breach of discipline.

I say, "Well Done."

A. A. Cumberledge
Captain, USN. (Ret.)
20 July 1995

(A brief biography of Captain Cumberledge is given
in Appendix A)

PREFACE

THE INCREDIBLE STORY of the U.S. Navy Expedition to Siberia, code name MOKO,[1] during World War II is about one small group of men who were involved in the last strenuous efforts to defeat Japan. Other isolated groups, particularly those in China, were exposed to far greater risks and endured more severe conditions,[2] but each was intent on providing weather information that would reduce casualties in the proposed invasion of Japan itself. Negotiations with the Soviet government for operating a weather base in Siberia represent the extreme example of how a bureaucracy based on fear can distort the character of an otherwise friendly people. There was great reluctance to reach a decision at the local or regional level because of fear of cruel reprisals for an apparently reasonable, but politically incorrect, solution to a problem. Even those already assigned to the grim Siberian environment had the threat over their heads of being sent to the notorious Sakhalin Island[3] based on decisions reached at the distant, detached, command center in Moscow. The Russian people encountered in Siberia, when divorced from the bureaucracy, appeared to be more like Americans than

---

[1] The letters in MOKO had no specific meaning. The advantage was that the letters were pronounced the same in Russian and English.

[2] A guerrilla unit, the Sino-American Cooperative Organization (SACO) was formed with Nationalist China to operate behind Japanese lines. Simplified weather equipment and radio transmitters were issued to a corps of weather observers protected by Maj. Gen. Tai Li's forces. As of 1 July 1945, the weather corps consisted of 34 officers, 57 aerographers and 8 radiomen (Stratton, 1950, pp. 107–108). A chief aerographer's mate with outstanding capabilities, Arthur N. Thomas, who had served with the author on the aircraft carrier USS Mission Bay (CVE 59), was assigned to SACO instead of the Siberian Expedition. One of the principal missions of SACO was to obtain sufficient information to make a weather chart for both free and occupied China and several hundred miles off shore (Bates and Fuller, 1986, pp.128–129). The most northern weather station reporting to the SACO Weather Central near Chungking was in Inner Mongolia at Shenpa (Hsiapachen) about 480 miles west of Peking (Peiping). Some of the exploits of SACO have been described by Stratton (1950) and by Miles (1967).

[3] The horrible treatment of prisoners on the Island of Sakhalin was described in great detail by the famous Russian master of the short story A. P. Chekhov in 1895 ("Ostrov Sakhalin," Moscow, 520 pp.). Several English translations are available. The island remains a symbol of the depths of human degradation.

xiii

any other people in the world. Their friendly, outgoing, pioneering spirit was matched by their rugged endurance, persistence, and hard work. One could roughhouse with them, make fun of their appearance, enjoy raucous jokes, and compete in friendly games—provided the internal security police were not watching. There was little doubt that in the absence of fear, Russians and Americans could become best friends.

The fear generated by the Soviet System was evident when people with whom you had been friendly in private the day before refused to speak to you in public. Associations and conversations with foreigners had to be reported, so any public contact with Americans was avoided. The absolute life-and-death control of officers over enlisted men in the army was witnessed directly—one jeep driver was saved from the death penalty for a simple error in direction by direct intervention of U.S. personnel. Needless to say, discipline was rigorously enforced, and every command received an instantaneous response of "I hear you." The following story celebrating the fiftieth anniversary of the U.S. Navy Expedition to Siberia during World War II is reconstructed in part from memory and the brief official history of principal events. Most of the more detailed official records were deliberately ordered to be destroyed, for purposes of security, by MOKO personnel on leaving Siberia. On the return trip by sea there was some time to make notes of the events experienced by the group for the historical record of the expedition for the U.S. Navy. The official history was not recoverable from the National Archives, because finding guides are not yet—and may never be—available for many of the records of CINCPAC (Commander in Chief of the Pacific Fleet). Because the author was involved in the writing of that history, the first draft of the preliminary notes was preserved. As a result of a request to Dr. Charles C. Bates (see Bates and Fuller, 1986) for use of a photograph in a completed draft of the manuscript, he mentioned that the actual source was the Archives of the Naval Weather Service at the Naval Meteorology and Oceanography Command at the Stennis Space Center in Mississippi. On inquiry at the Archives in Mississippi, Cmdr. Thomas V. Fredian, with the assistance of Mrs. Cathy Willis, of the Public Affairs Office was able to provide not only a copy of an onion skin copy of the official history of the MOKO Expedition, but also twenty of the conference reports filed by the commanding officer! Cmdr. Fredian had personally rescued the history from destruction by a newly arrived senior officer who had little regard for "old" records then only a few years old. We were able to confirm the official account of events that were described in the first draft notes, which were found to be accurate. In addition, the author's personal flight log book required of an Air Observer has been preserved all these years along with other minor documents that were allowed to pass out of the USSR.

More than half of the nineteen officers remain alive today, but little success was achieved in locating the fine group of forty-one enlisted men who served with us. A notice in the newsletter (*The Aerograph*) of the Naval Weather Service Association yielded only one response from an officer. For this reason, documentation of the more personal events of the enlisted men is minimal, and it is hoped that an accurate accounting has been made. The manuscript has been reviewed by seven officers on the MOKO Expedition: Charles E. Birkett, Charles L. Bristor, the late Arthur A. Cumberledge, the late Frederick A. Nelson, Parke P. Starke, Byron Uskievich, and Philip Worchel, who provided additional comments from their own recollections, which are specifically identified accordingly. Where the statements in previously published general accounts by others not on the expedition are in conflict, preference was given to the "facts" as described in the official history and to the observations of the currently living officers of the MOKO Unit. Perhaps the author can be forgiven if some minor, personal events are mixed up in regard to place and timing, but most of the actions remain as clear as if they had occurred yesterday. It was an experience that the entire group no doubt recalls vividly whenever the word Siberia is spoken.

I had served as an aerologist (meteorologist) at several Fleet Weather Centrals, an Air Wing involving flying as a technical observer, and had one and a half-years of sea duty on aircraft carriers before assignment to the MOKO Expedition. My meteorological training was obtained at the University of Chicago under the internationally famous Prof.

Carl-Gustav Rossby.[4] In addition, I had the good fortune of serving under Captain Wilbert M. Lockhart, dean of the weather central concept in the Navy Department, at the San Francisco Weather Central after completing the intensive training program at Chicago. That I was originally trained as a geologist will become evident from some of the events described below.

None of the opinions expressed herein are to be attributed in any way to the U.S. Navy or any other branch of the armed services.

---

[4] The training program in meteorology at the University of Chicago was initiated in the spring of 1940 by Dr. Horace R. Byers, then in the U.S. Weather Bureau at the Chicago District Forecast Center. The first classes began in the fall of that year (Byers, 1976). Byers persuaded the University to appoint Dr. Carl-Gustav Rossby as chairman of the new Institute of Meteorology under the Department of Physics, and he arrived the next year after a year's leave at the Weather Bureau in Washington, DC. At that time there were only three graduate programs in meteorology in the U.S.: California Institute of Technology, Massachusetts Institute of Technology, and New York University. Including the Army Air Corps Technical Training Centers at Chanute Field, IL, and Grand Rapids, MI, the Naval Postgraduate School, Annapolis, MD, as well as ten other universities, about 7,000 weather officers were trained in a brief three-year period (Byers, 1970).

1. INTRODUCTION

Weather—a critical factor.

THE INVASION OF JAPAN was planned by the Joint Chiefs of Staff to take place in November 1945, initially with the first assault on the island of Kyushu and subsequently on Honshu in March 1946. Because the attack was to be launched primarily by carrier-based aircraft and B-29 bombers from the Marianas, knowledge of the weather over the mountainous Japanese islands was essential for the successful preparation of landing areas for ground troops. The weather systems generally move over Japan from the northwest, that is, from Siberia, and travel mainly from west to east in these latitudes of "prevailing westerly winds." The cold Siberian air crosses the Kuroshio current (Fig. 1) and produces well developed polar fronts with strong thunderstorms. As a result of high velocity jet streams, bombers were literally stopped in their tracks and caused serious navigational and fuel consumption problems. The existing weather stations in China were too far south and well inland to be of help to the U.S. Naval forces. For these reasons, it was essential to have weather stations up-wind from Japan if there was to be reasonable success in predicting the weather over Japan. The Fleet Weather Central in Pearl Harbor, HI, was initially responsible for tracking the typhoons that originated in the vicinity of the Marshall Islands, moving west toward the Philippines or curving northwest toward Japan. Preliminary compilations of typhoon tracks (Fig. 2), assembled by the U.S. Weather Bureau from ship's weather observations, were available from the U.S. Hydrographic Office for the critical months preceding and during the planned invasion (Anonymous, 1943). The typhoons of December 1944, and June 1945 had disastrous effects on Adm. William F. Halsey's Third Fleet[5]; there-

---

[5] On 18 December 1944, Typhoon Cobra caught Halsey's fleet about 300 miles east of Luzon. Three destroyers went down with practically all hands, serious damage was inflicted on 9 ships, and at least 19 others sustained lesser damage. About 146 aircraft on the various carriers were lost by fire, smashed, or swept overboard. Worst of all was the loss of 790 officers and men and some 80 injured. Typhoon Viper on 5 June, 1945, caught 48 ships of the Fleet southeast of Oki-

1

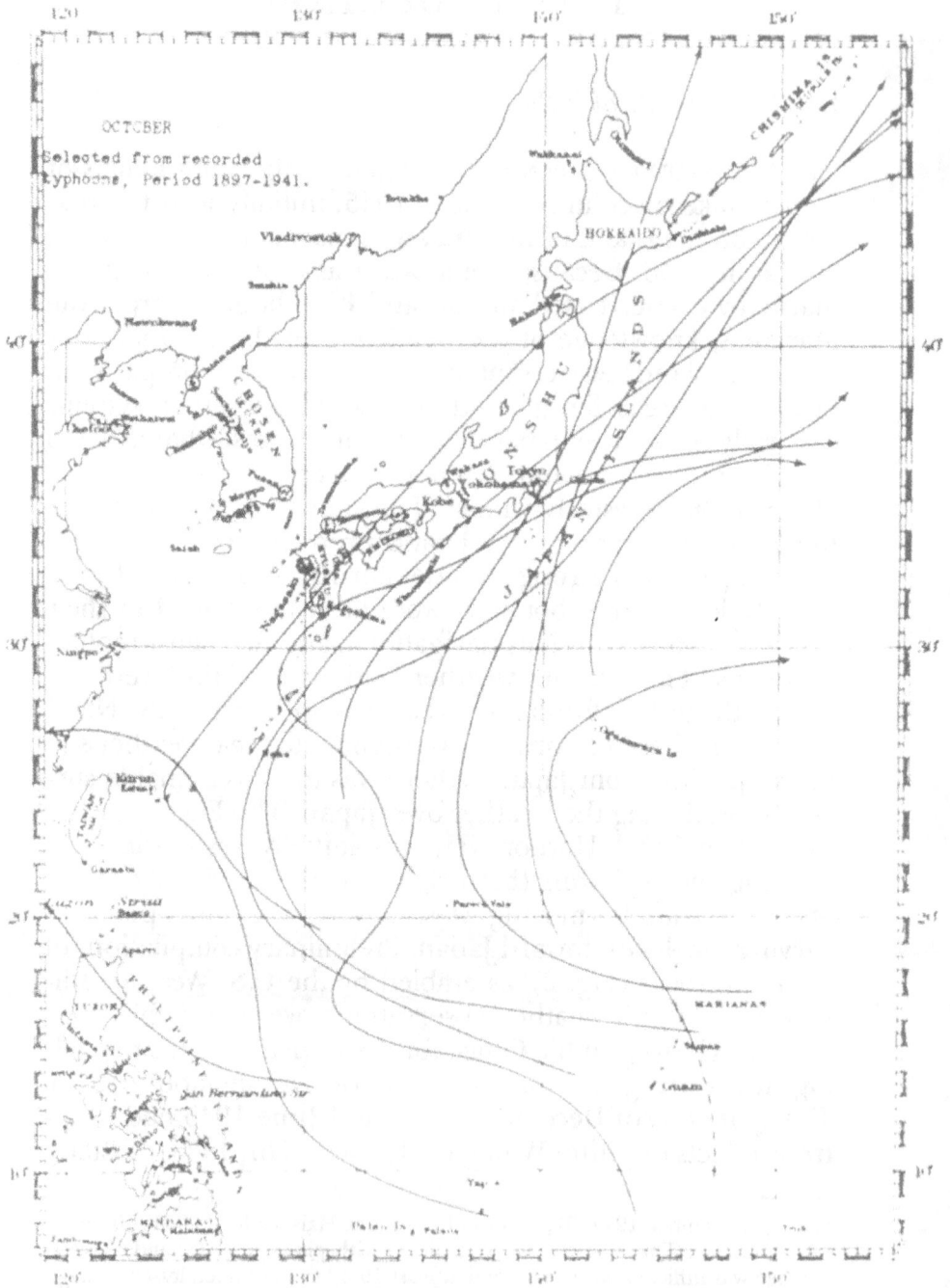

Fig. 2. Compilation of recorded typhoons for the month of October, 1897–1941. U.S. Hydrographic Office, 1943.

after, airborne weather reconnaissance flights tracked these intense storms. Eventually, a special forecasting unit for typhoons was set up closer to their tracks, at the Fleet Weather Central in Guam, to advise the Commander in Chief of the Pacific Fleet, Adm. Chester W. Nimitz, of the weather and sea conditions around Japan. In November 1945, the risk of typhoons in the area southwest of Japan was minimal, but, nevertheless, a hazard to be given very serious attention for the staging areas of the invasion (e.g., the typhoon over Okinawa on 9 October 1945, would have had disastrous effects on troop transport). The great contrast in weather over Japan is exhibited in relevant portions of the World Climatic Charts for July (Fig. 3) and January (Fig. 4) published in July, 1943, by the Naval Meteorological Branch of the Admiralty in London and made available to the U.S. Navy. Because it was anticipated that over 100 aircraft carriers [28 first line carriers (CV's) and 71 escort carriers (CVE's)[6] under the U.S. flag were then available] would take part in the initial attack, a 24-hour advanced warning of inclement weather was a critical factor in those military operations.

It was recognized as early as 1941 that naval operations in the Pacific would require weather information from vast areas that were essentially inaccessible to the U.S. During the Harriman–Beaverbrook Mission to the USSR[7], a weather exchange program was initiated with the Soviet government.

---

nawa, and the eye of the storm passed over them causing serious damage to four vessels (Adamson and Kosco, 1967).

[6] The number of carriers is estimated from the statistics for 14 August 1945 (VJ Day) in the unpublished summary of "U.S. Navy Ship Force Levels, 1917–1989" prepared on 8 December 1989 by the Ships History Branch, Naval Historical Center, Washington, DC.

[7] The Harriman-Beaverbrook mission to Moscow began its three-day deliberations with Stalin on 28 September 1941. The Germans had attacked the Soviet Union on 22 June along three fronts sweeping over an area twice the size of France. Although Stalin wanted a second front "somewhere in the Balkans or France" he also asked for massive supplies of armament. Harriman raised the question of delivering aircraft via Alaska through Siberia. Stalin was anxious to get the airplanes but said it was "too dangerous a route" for American crews to ferry the planes. Furthermore, he did not want to jeopardize his neutrality pact with Japan, but admitted that he had earlier sent airplanes and artillery to Chiang Kai-shek in China to aid in their fight against Japan. Stalin dismissed the conflict in principle on the grounds that there were no provisions in the treaty with Japan against it!

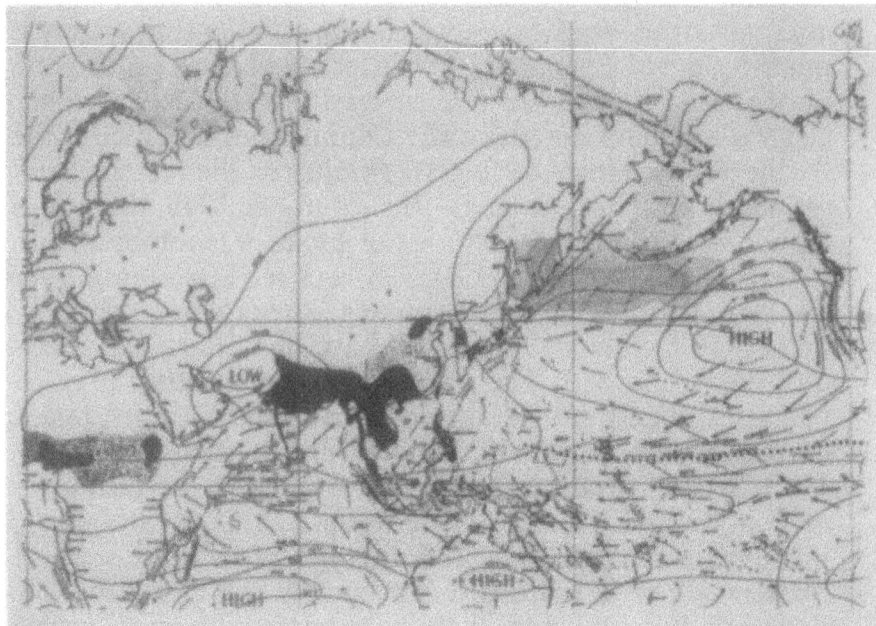

Fig. 3. Portion of the World Climatic Chart No. 5302, Sheet 2, for July. Published by the Admiralty, 16 July 1943. The principal feature is the southeasterly flow of air over the Japanese Islands. The dotted areas denote sea fog and the heavy shaded areas indicate high rainfall.

In return for Lend-Lease supplies, the Soviets acquiesced to an exchange of data between Khabarovsk, Siberia, and San Francisco, CA, from about 15–20 weather stations. Unfortunately, the data from the Siberian stations essentially proved useless for several reasons that will become clear below. After an extended visit to the U.S. by a mission from the Soviet Hydrometeorological Service in February, 1943, the number of stations that were to supply data from Siberia was increased to 30. Further expansion of the weather exchange program resulted in some benefit to Allied European operations, but the meteorological information coming out of Siberia continued to be marginal even though described in politically euphemistic terms as "extremely valuable."

Agreement—in principle.

Aside from the general reluctance of the Soviets to give the U.S. any information because our intentions were viewed

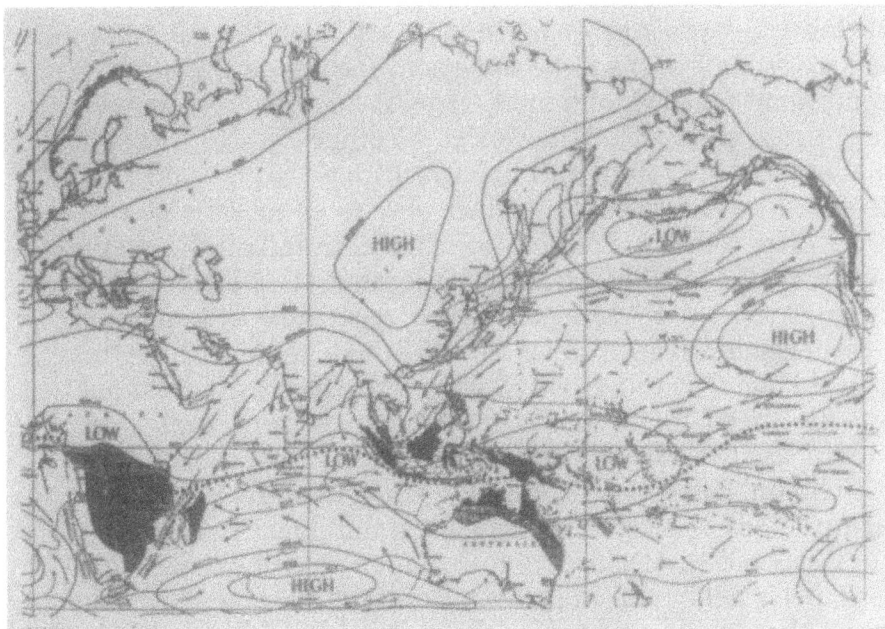

Fig. 4. Portion of the World Climactic Chart No. 5301, Sheet 1, for January. Published by the Admiralty, 16 July 1943. The principal feature is the cold dome of high pressure over Siberia and the northwesterly flow of air over the Japanese Islands. The brackets outline the limit of sea ice.

with suspicion,[8] the U.S. weathermen believed that perhaps the equipment being used in the vast stretches of Siberia was inferior and the data being collected were just not compatible with that obtained with U.S. instruments. On these grounds, an offer was made at the Yalta Conference in February 1945 to equip and man 37 weather stations in Siberia. Again, the Soviets agreed to receiving the equipment,[9] but balked at

---

[8] The suspicion was mutual from the very beginning. Churchill had pointed out to the Soviet Ambassador in London as early as September 1941 that only four months before it was not clear whether the Soviets were going to be against us on the German side. In fact Churchill had been persuaded that they would join the Germans, and was angered by the sudden demand for a second front.

[9] The equipment was actually shipped to Murmansk in June and July of 1945 after many delays in obtaining authorization from the Soviet Purchasing Commission and assignment of a cargo space on available ships. Lieutenant General Eugene K. Fedorov acknowledged on 27 August 1945 that the equipment had been received and was being installed. He apparently was pleased to have the

having U.S. personnel man the stations. Even the attempt to send U.S. staff to train the Soviets in the use of the modern equipment failed. Lieutenant General Eugene K. Fedorov, a scientist and Chief of the Soviet Weather Service, thought manuals on operation and maintenance would be sufficient. In short, no American personnel were to enter Siberia.

After VE day, attention was concentrated on the events in the Pacific. At the Potsdam Conference (17 July–2 August 1945),[10] the issue was again raised regarding U.S.-manned weather stations in Siberia. Even though weather bulletins were being received from Khabarovsk, the data were poor, and the need for accurate information became ever more urgent as invasion plans were developed. After consulting with Stalin, General A. I. Antonov, Chief of Staff of the Soviet Army, finally approved in principle a minimal plan. The Soviets had not yet declared war on Japan, and the delays may have been deliberate in order to forestall any compromising move that would jeopardize their defense in the west. The entry of U.S. forces into Siberia could be interpreted by Japan only as a hostile act. On the other hand, the Soviets wanted to be in on the defeat of Japan and share in the spoils.[11] After much heated debate and direct pressure from President Harry S. Truman, the U.S. finally got permission on 26 July 1945 to set up two Navy weather stations, manned by U.S. personnel with their own communication facilities. One was to be positioned at Petropavlovsk, Kamchatka Peninsula, and another at Khabarovsk on the Amur River.[12] These stations were purposely designed in no way to duplicate the Soviet weather services. Thus, the specialized bulletins and analyses would be in a form readily recognized and usable to the U.S.

---

new equipment even though the Sino-Soviet Peace Treaty had been signed on 14 August 1945.

[10] The Berlin (Potsdam) Conference was held from 17 July to 2 August 1945 and was attended by the USSR, U.S. and U.K.

[11] During the Yalta Conference held 4–11 February, 1945, it had already been agreed that the southern part of Sakhalin, which had been lost to Japan in 1905 as a result of the Russo-Japanese War, would be returned to the Soviet Union.

[12] There is no record in the State Department of a formal agreement regarding the establishment of weather stations in Siberia. The detailed list of Lend-Lease materials requested does not include equipment for weather stations. Petropavlovsk had been used as an emergency landing field, but the personnel forced to land there were interned.

Armed forces. It would certainly eliminate the difficulties of copying and interpreting foreign broadcasts.

Invasion alternative.

Events thereafter moved at incredible speed: the first atomic bomb (based on uranium 235) was dropped on Hiroshima on the island of Honshu on 6 August 1945 while the leaders of the Amphibious Forces were meeting aboard a flagship in Manila Bay to plan for the final assault on Japan. Among the solutions for achieving the mandated "unconditional surrender,"[13] the dropping of an atomic bomb was a political long shot. It was unique and not comparable to the classical military alternatives: 1) naval blockade, to cut off supplies and starve out the enemy; 2) aerial bombardment, to destroy their industrial means for war; and 3) amphibious assault and massive troop invasion, to neutralize their fighting forces. Japanese fanaticism and suicidal defense techniques using kamikazes as well as the natural impenetrable cave defenses, already demonstrated on Okinawa and Iwo Jima, emphasized the need for an integrated solution. Each alternative had its proponents, but early planning indicated that all approaches were needed for the defeat of such a strong-willed and disciplined enemy. For examples, aerial mining of the strategic harbors and straits by the Army Air Force was needed as a supplement to the Navy blockade; precision daylight bombing of industrial areas by long-range bombers required acquisition of bases by the Army and Marines within range of the targets; and the army needed short-range fighter protection for the armada of transports and landing craft to be used in an invasion. As early as 3 April 1945 the order had come down from the Joint Chiefs of Staff to begin planning for the invasion of Kyushu (operation Olympic) and of Honshu (operation Coronet) (see Skates, 1994). The invasion plans were approved by President Harry S. Truman on 18 June 1945 (minutes of meeting with the Joint Chiefs of

---

[13] The concept of "unconditional surrender" was outlined by President Franklin D. Roosevelt on 7 January 1943 at a meeting for the Joint Chiefs of Staff and announced at the end of the Casablanca Conference with the British on 23 January 1943 (see Armstrong, 1961).

Fig. 5. Japanese stamps issued on 1 May 1945 that referred to the planned invasion of Japan by Kublai Khan in 1281. All three are 10 Sen, typeset (=letterpress) with curved wavy line watermarks. Left: Scott # 335, light gray, 13 perf (13 perforations per 2 centimeters). Center: Scott # 354, imperforate, light gray. Right: Scott # 354A, imperforate, blue. (Author's collection).

Staff) after detailed consideration of estimated potential casualties.[14] The invasion would constitute the greatest amphibious assault of all time involving 760,000 troops, about 2,500 ships, and some 5,000 aircraft.

Despite the successes of the Allied Forces, the Japanese officially issued postage stamps on 1 May 1945 indicating the country's intention to resist any invasion of the homeland. Three stamps (Fig. 5) demanding the surrender of the "Enemy Country," purposely referred to an event in 1281 regarding the planned invasion of Japan by Kublai Khan (Winick,

[14] The discussion of casualties by the Joint Chiefs of Staff was extensive after reaching the conclusion that invasion of Kyushu was necessary. Adm. Leahy pointed out that the troops in Okinawa had lost 35 percent in casualties. If the assault troops in the Kyushu campaign numbered 766,700 as determined by General Marshall, then the casualties in the first 30 days would be in the order of 268,000! Adm. King, on the other hand, thought that a more realistic number would "lie somewhere between the number experienced by General MacArthur in the operations of Luzon and the Okinawa casualties." If the same number of troops were used in Luzon and Okinawa, the average number of casualties would be 30 percent, or 230,000. (In comparison, 42,000 were lost by the U.S. in the

Fig. 6. Industrial buildings in Nagasaki, Japan, obliterated by atomic bombs dropped on 9 August 1945. Photograph was taken by unknown U.S. Army officer.

1996). The Mongolian invasion fleet was destroyed for the most part by a typhoon[15]—a lesson that was nearly repeated in 1945.

Soviet participation.

The Soviets declared war on Japan on 9 August 1945, the same day that the industrial area of Nagasaki, island of Kyushu, was obliterated by a second atomic bomb (plutonium-based) (Figs. 6 and 7). The advantages to be derived from Soviet entry into the war with Japan was first suggested by President Franklin D. Roosevelt to Soviet Ambassador M. M.

---

first 30 days of the Normandy Invasion.) The attrition of the Japanese, killed and taken prisoner, was expected to be from 2 to 5 times those of the U.S. Delay in the invasion only favored the enemy, and the Joint Chiefs of Staff agreed with the President that after "considering all possible alternative plans . . . the Kyushu operation was the best solution under the circumstances." (Anonymous, 1955, pp. 79–83).

[15] That typhoon was described by the Japanese as the "divine wind," or kamikaze—a word that was applied in WW II to the pilot of a bomb-ladened aircraft whose sole mission was a suicidal crash dive on his target.

Fig. 7. View of destruction of Nagasaki, Japan, from atomic bomb.
Photo taken in December 1945, by unknown U.S. Army Officer. It is not
known if the shrine-like portal entrance (Tojii) was erected before or
after the blast.

Litvinov on 8 December 1941, the day after the attack on
Pearl Harbor, HI. The U.S. was officially informed on 11 December 1941 that the Soviet Union was not then in a position
to cooperate in operations against Japan because of the risk
of attack by Japan in the east. At that time they were "protected" by the Russo-Japanese Neutrality Treaty signed on 13
April 1941. Needless to say, it gave the Soviets an opportunity,
albeit tenuous, to survive the imminent onslaught of the
Germans in the west in June 1941. In November, 1943, Stalin
had given a verbal commitment at the Teheran Conference
to enter the war against Japan. Assurances were given at the
Moscow Conference on 17 October 1944 that the Manchurian offensive was to begin three months after the defeat of
Germany. These assurances were reaffirmed at the Yalta
Conference in February 1945. The planners thought the Soviets could defeat, or at least tie up, the Japanese Army in
Manchuria, a force of some 713,000 men, so it could not reinforce the home defenses in Japan against invasion. In addi-

tion, the Soviets could provide air bases in Siberia to support shuttle bombing[16] of Japan. The Soviets could also disrupt Japanese shipping of supplies to the Mainland. To these contributions could be added the strong psychological impact of the entry of an historical enemy, the USSR (Russo-Japanese War of 8 February 1904–5 September 1905), against the Japanese people. The Japanese ambassador to the USSR was nevertheless surprised at the Soviet declaration of war against Japan on 9 August 1945, particularly in view of their solicitations to the Soviets for them to act as mediators in restoring peace with the signers of the surrender proclamation.

The anticipated contributions of the Soviets were rapidly dismissed by the time of the Potsdam Conference in July 1945. The Japanese Army in Manchuria had been drastically weakened by poorly-trained replacements. There was no longer a need for air bases in Siberia because the long-range B-29s could reach targets in Japan from the Marianas. Japanese shipping had been decimated by mine laying and naval engagements. By mid-June 1945, 66 cities had been burned out and 330,000 Japanese were incinerated (Hansell, 1980). Furthermore, the shock of the fire-bomb raids was having the desired psychological effect on the people even though some of the planners did not believe it was a decisive strategy. It was not obvious that the opening of a northern Pacific sea route to deliver the enormous amount of supplies required by the USSR from the U.S. was feasible. Another concern was the cutting of the Trans-Siberian Railroad and the occupation of the Vladivostok port by the massive Manchurian Army. In short, the planners no longer believed Soviet entry into the war was essential for the success of the invasion of Japan (see Anonymous, 1955). Nevertheless, a meteorological base in Siberia was critical for forecasting the weather conditions for the invasion.

The entry of the Soviets into the war with Japan on 9 August 1945 is viewed by some as the result of a quick de-

---

[16] Shuttle bombing is a round-trip flight technique whereby bombs are loaded at an advanced airfield, dropped over a designated enemy target, and the flight continues on over enemy territory to a safe allied airfield. Planes are refueled there and another load of bombs installed. The flight bombs the same or other designated enemy targets on the return run to the home base. The Army Air Forces used the airfield at Poltava in the Ukraine for this purpose (See Postscript).

cision to take advantage of an opportunity to reap benefits for the USSR. It is also likely that Stalin was living up to his promise of beginning the Manchurian campaign three months—almost to the day—after the defeat of Germany. Because of the slow process of decision making in the Soviet bureaucracy and the long lead time to prepare an offensive, one must give the Soviets credit for fulfilling that promise. On the other hand, the promise of air bases in Siberia was never fulfilled even though U.S. teams for airfield development sat for weeks in Fairbanks awaiting visas that never came. On the grounds of maintaining secrecy, the Soviets even expected the U.S. teams to wear Russian uniforms (Deane, 1946, pp. 252–253).

The entry of the Soviets into the war against Japan was not without a political price. In addition to the mountain of supplies to be delivered by the U.S. across a northern Pacific sea route yet to be opened and maintained, the Soviets wanted the return of the southern part of the Island of Sakhalin, annexation of the Kurile Islands, restoration of their holdings in Manchuria, recognition of the *status quo* in Outer Mongolia, and the internationalization of the port of Dairen (Anonymous, 1955, p. 46). Marshall N. N. Voronov (1965, pp. 17–18) negated some of the perceived opportunism when he wrote:

> "He [Stalin] told us [Marshalls Voronov and Sokolovski] that we should make use of the favorable international situation to return everything that Japan had seized as a result of the Russo-Japanese War." He added: "But we don't need anything that isn't ours." This was a great military secret which wasn't supposed to be mentioned to anyone, but at the same time we had to be very active in a quiet way.

Apparently the Soviet troops were not aware of the decision to begin moving troops and supplies to the East while the final battles were still being fought in the West.

Orders cut.

While these momentous events were underway, orders were cut (issued) on 4 August 1945, to assemble the personnel for the two weather stations. On 9 August 1945, the orders were explicit and urgent to report via air with a class one priority to Seattle, WA, for further transportation to Naval Advanced Base Unit MOKO.

The Commanding Officer Capt. A. A. Cumberledge, then serving in the Office of the Chief of Naval Operations, was selected by Capt. Howard T. Orville, Chief of the Naval Aerological Service and approved by Adm. Ernest J. King. Capt. Cumberledge had the privilege of selecting his own Executive Officer, Cmdr. E. E. Butow. The other weather officers and men were assigned by the Detail Officer, Cmdr. Richard Steere, in Washington from the list of 1,318 aerological officers and some 5,000 enlisted aerographers.[17] The support staff was selected on the basis of skills needed by the Expedition and availability by the specialty detailers in the Bureau of Naval Personnel.

---

[17] The number of aerological personnel available at the beginning of World War II was only 90 officers and 600 enlisted men (Anonymous, 1957).

2. ASSEMBLY OF THE MOKO UNIT

Destination revealed.

THE LAST OF THE VARIOUS GROUPS of officers and men destined for the MOKO Unit did not arrive in Seattle until 1 P.M. on 16 August because of engine trouble and other delays precipitated by the scheduling of the Naval Air Transport Service. In the meantime the President of the U.S. announced at 7 P.M. on 14 August 1945 that the Japanese had accepted the terms of unconditional surrender outlined at the Potsdam Conference.[18] The Soviets reaffirmed on 17 August 1945 all previous arrangements regarding the establishment of weather stations in Siberia in spite of the unconditional surrender of the Japanese. It was certainly not evident to either the USSR or the U.S. that all Japanese would surrender peacefully. Furthermore, the need for accurate weather information was not diminished by the enormous task of transporting troops to Japan by sea and air, mine-sweeping operations, and relocation of the Pacific fleet. Departure of the MOKO group for Siberia was scheduled for noon on 20 August 1945.

The group established the Seattle Weather Central as headquarters, and the Commanding Officer (CO) concluded that six days would be required to finish preparations. Immunizations were completed. The Ship's Service Shop, Uniform Shop, Small Stores, and the availability of a major supply depot facilitated the outfitting of the personnel except for Arctic gear.

At last, during the morning meeting of officers and men on 18 August 1945, the crew was informed of the destination and general duties. All fears of being sent to a remote jungle outpost evaporated.[19] Some of the crew had volun-

---

[18] A proclamation was issued on 26 July 1945 defining the terms for Japanese surrender by China, U.S., and U.K. It was not signed by the USSR.

[19] Regular officers of the Navy advised reservists to "Never, never, never, request shore duty when serving at sea." A clean, comfortable ship surpasses any shore-duty station outside the U.S. There were indeed some unpleasant posts around the world that invite rotation of personnel, so it was wise not to complain—no matter how intolerable the conditions were perceived to be on shipboard.

Fig. 8. Flat-bed truck carrying twelve pieces of fire-fighting apparatus delivered as a result of a typo error on the order for hand-held fire extinguishers.

teered on the basis of inside information, but had kept the secret. Others wondered what they had done wrong to warrant being sent out to such a remote and hazardous place. The welcome presence of four Russian-speaking officers and two fresh out of language school was revealed. After that announcement, assignments were passed out. Some 60 tons of equipment needed for a six-month mission had to be collected. All were cautioned about the TOP SECRET nature of the mission. Planes were made available to officers who were unable to fulfill their want lists locally; these were flown directly to the manufacturer when necessary to obtain the required materials. In the rush, some errors were made, but the most glaring was the arrival of a large, flat-bed truck bearing twelve pieces of fire-fighting apparatus on wheels (Fig. 8). A typographical error in a digit of the code number describing the type of apparatus on the order forms for twelve small hand-held fire extinguishers was quickly corrected. After around-the-clock efforts, most of the gear was acquired in the allotted time, and the first group was in the air on time on their 6-hour flight to the first stop at Fairbanks, AK. Because of weight limitations, an officer was detailed to handle the shipping home of any unnecessary personal gear still in the bags of the men. Another officer handled all the last-minute personal affairs of the first group departing. Three large transport aircraft, R5D-3s,[20] operated by Squadron VR-5, commanded by Cmdr. Kenneth Haigens, were made available to transfer the group and its equipment. It was assumed three round-trip flights would be needed to complete the transfer, but three proved to be insufficient.

The headline of the local newspaper issued on 20 August 1945 in Fairbanks (Fig.9), when the first group arrived, was obviously viewed with surprise by members of the Unit. The occupation of Japan was clearly not expected to be free of incidents. Even the "Tokyo newspapers cautioned that 'control of the military' would be the major problem." The Japanese army, still intact and undefeated, was capable of resisting any allied invasion or occupation despite the wishes of the emperor. Negotiations with individual Japanese field units were underway in most occupied areas. Virtually all the fighting in Manchuria had ceased, but a month or more was considered necessary to achieve full surrender. [One Japanese officer in the Philippines, still in uniform and armed, did not surrender until 1974 (Anonymous, 1989)!] The landings on the home islands would no doubt have a stunning impact on the Japanese, and the response was unpredictable.

Another item on the front page of the same newspaper may have influenced the subsequent lack of cooperation by the Soviets. President Truman had announced that Lend-Lease shipments were to be terminated on 20 August 1945.

---

[20] The R5D-3 is the Navy designation for the four-engine commercial transport aircraft, better known as the DC-4, made by the Douglas Aircraft Company. (The U.S. Army Air Force labeled it C-54.) The three planes (Bureau Nos. 56530, 56531, 56532) were delivered new on 13 August 1945 to Oakland, CA. A normal crew of five operated the aircraft. The basic cargo payload was about 13 tons with a cruising speed of 210 mph and a range of 2,290 miles (Swanborough and Bowers, 1968).

It was reported in Craven and Cate (1983, p. 733) that "B-29s carried equipment to Khabarovsk in Siberia in a belated effort to set up a weather station. . . ." The operations reports (DC/S A-3, Project Reports 1–3 for 18–20 August 1945) that were cited from the 20th Air Force were accessible from the Air Force History Support Office, Bolling AFB, Washington, DC. The mission to transport weather equipment from Manila and Okinawa was terminated and not executed according to historian Dr. Walton S. Moody. In addition, the combat chronology of the Army Air Forces (Carter and Mueller, 1973), does not list operations for those dates. To the best of my knowledge, no weather equipment was transported by the Army Air Forces to Khabarovsk. [The meteorological equipment for Soviet weather stations was sent via ship to Murmansk (see footnote 9).]

Fig. 9. Portion of the front page of the Fairbanks Daily News-Miner published on 20 August 1945 in Fairbanks, AK.

Delaying tactics.

In Fairbanks, the hassle with the Soviets began with a full head of steam. First, they insisted on a Soviet navigator in each plane, a Russian-speaking translator to deal with airport communications, and all U.S. military personnel were to have passports and visas! The next request was for all officers and men to abandon their uniforms and proceed as civilians. There were to be no weapons carried. And those demands were just the beginning. Every effort was made to persuade the Soviets that we were allies and uniforms were a necessary protection under the Geneva Convention. A group visa was arranged, the presence of a Soviet navigator and translator was accepted, as was the ban on weapons. At Soviet insistence, all U.S. literature was to be left behind and no cameras were permitted. Although the general in charge of the Soviet detachment in Fairbanks claimed he had no previous knowledge of the U.S. mission to Siberia, he just happened to have navigators and radiomen standing by for an "unknown purpose." The often to be used ploy of sending a dispatch to Moscow for instructions was the Soviet general's next move.

While these delays were being promoted and compounded, the remaining two planes arrived. Care had been taken to divide up the men and equipment so that the loss of any one plane would not result in scrubbing the mission. The safe arrival of the three aircraft in Fairbanks was not given a second thought by the aircraft crew or their passengers mainly because they were not aware of the risks endured by their predecessors. As weather officer on the maiden and a few subsequent nonstop flights between Seattle and Kodiak, I appreciated the hazards of the over-water route, but reasoned that it could be managed successfully if one put his meteorological knowledge to immediate practical use. The serious effects of weather as well as mechanical problems on the early flying operations to Alaska was summarized in a 1942 report by an air force officer quoted by Cohen (1981, p. 35):

> Early in January, twenty-five P-40s of the 11th Pursuit Squadron set out for Elmendorf Field (Anchorage, Alaska). On January 12, twenty-two were still en route, three having crashed before reaching the Canadian border. Ten planes left the

United States and were somewhere between the Canadian border and Ladd Field (Fairbanks, Alaska). Nine were in Spokane
and three were at Portland. Of the five transports that were to
accompany the pursuit planes, three were still in Spokane and
two were somewhere in Canada. Thirteen B-26s of the 77th
Bombardment Squadron also set out for Alaska at the same
time as the pursuit planes. On January 12, nine of them were
somewhere between Spokane and Ladd Field and four were
still in Spokane. Almost a month later, only thirteen of the P-
40Es had arrived at Elmendorf Field. Seven had cracked up
en route and only four of these were salvageable. Five were
still on the way. The B-26s met with even greater difficulties.
Out of the thirteen that started for Alaska, five crashed. Four
of the crackups took place in what is now known as "Million
Dollar Valley" between Edmonton and Fairbanks. Of the eight
that finally arrived, four were grounded at Fairbanks and four
at Elmendorf with faulty fuel installations.

The over-water route clearly had superior advantages
to the over-land route to Alaska, but it was just as well the
passengers were not aware of the comparable hazards. For
example, after ditching in cold seawater, a man in wet clothing has only about 1.5 hours of survival time if the temperature of the water is 0° C (32° F); 2 hours at 5° C (41° F); 3.5
hours at 10° C (50° F); 6 hours at 15° C (59° F); and might
last as long as 18 hours in seawater that is 20° C (68° F) according to Molnar (1946).[21]
Arctic gear was issued from the U.S. Army Quartermas-
ter stocks; it was indeed a substantial collection of apparel
designed and tested to keep one warm. From fur hats to
mukluks,[22] face masks and goggles to strap-on ice prongs,
head socks to foot socks, mufflers to heavy-wool long underwear, as well as an eider-down sleeping bag were issued in

---

[21] These values will vary according to an individual's nutritional and hydration
status, shivering capacity, activity, state of fatigue, injuries suffered, as well as the
number of layers of clothing and exposure to wind. The limits are derived from
the records of shipwreck survivors.
[22] The mukluk is a well-designed boot patterned after the sealskin Eskimo boot.
The rubber bottom with felt liner is especially protective in wet cold. The boot
has a laced leather top that is flexible for climbing yet holds knee-high socks in
place. Leather flying boots lined with sheep skin were also issued but were too
cumbersome for heavy outdoor work. They served as very comfortable slippers
at night indoors.

each man's gear. The good advice of keeping your head and belly warm helped, but when the temperature dropped below −40° F nothing seemed to be adequate. On 21 August 1945 all officers, men, and flight personnel were briefed on the customs and regulations of the USSR by two of the officers in the group who had past experience in the country. Top security was again emphasized, and the commanding officer cautioned that under no circumstance were any actions to be taken that could even remotely be interpreted as espionage. The primary mission was to provide accurate weather information to the Navy—no more, no less.

After a delay of about sixty hours, the commanding officer of MOKO thought that sufficient time had elapsed for a reply from Moscow and appealed to the Chief of Naval Operations for instructions. He was informed that complete clearance had been obtained in Washington on 22 August 1945 and he was to proceed with the mission. It may have been a coincidence, but on the same day the Soviets abandoned their plan to invade Hokkaido, Japan, from southern Sakhalin (Karafuto), which they had almost completely occupied. Extricating the Soviets from Hokkaido after the termination of the war might have precipitated another major political confrontation. The Soviet General agreed in Fairbanks to a meeting at 1:00 P.M. on 23 August at which time he said he would have definite information from his superiors. Upon learning that the time of take-off had already been set, the proposed meeting was canceled. A spotty briefing was arranged and the flight crews met with the Soviet navigators and radiomen to determine the route and radio frequencies to be guarded. Weather information had been exchanged[23] between Nome and Velkal, the proposed landing site in Siberia. The planes took off at 3:00 P.M. on 23 August 1945. The

---

[23] The weather data exchange had apparently been approved at a meeting on 17 May 1945 between the U.S. Army Weather Squadron Commander and Commander Ryzhkov of the Soviet Hydrometeorological office at Ladd Field, Fairbanks, AK. The uncoded data were transmitted on a low-frequency transmitter supplied to the Soviets by the U.S. (Minutes of the meeting at Ladd Field, Alaska, were taken by Warrant Officer (jg) Robert E. Most, U. S. Army. They are available at the Air Weather Service, Scott Air Force Base, Illinois.)

group was treated to a spectacular view of the intricately braided streams[24] of the Tanana and Yukon Rivers until poor weather was encountered over Nome and the Bering Sea. The weather was excellent, however, at Velkal when the planes landed a little over five hours later. The copilot entertained us en route by demonstrating the great sucking force of the air passing a small open porthole. The remains of the box lunches were disposed of by holding the boxes several feet from the opening. They were snatched out of the hand, deformed to pass through the porthole, and passed over the tail surfaces in the air stream of the airplane. It was a simple demonstration of the Venturi principle whereby the pressure is reduced inside the cabin with an increase in velocity of air flowing over an opening.

Velkal—more delay.

The airstrip at Velkal was presumably an emergency alternate landing place for the fighter planes being ferried by the Soviets from Fairbanks to the Soviet western front in the event of bad weather at Anadyrsk or Markova. It was a single strip parallel to the beach on the Gulf of Kresta. The 4,000-foot runway was made of 2″ × 12″ planks about six feet long set on edge and overlapping only a few inches. The space between planks was filled with gravel. It was very noisy—like a wash-board road—but held a heavily loaded airplane in all types of weather. The shoulders of the field were of soft glacial till—as one plane discovered while taxiing. Several U.S.-built fighters (Mustangs, P-40s) and the two-engine Douglas DC-3s were parked nearby. All of the equipment in sight (e.g., gas trucks, jeeps, tractors) for use in the ferrying operation were built in the U.S.

---

[24] Braided streams are especially common in glaciated regions. Glacial melt water may carry heavy loads of sediment that is deposited in their channel at low water, during excessive evaporation, or by seepage into their beds. The stream may then overflow its channel, cut a new channel, which in turn becomes rapidly choked. The process, termed braiding, develops a network of channels on the valley floor with successive branching and eventually rejoining the main channel. The tangled network of interlacing channels that separate and merge is described as a braided drainage pattern.

Fig. 10. Sod hut in a forested area of northern Siberia similar to those seen on the tundra of eastern Siberia. This particular hut was alleged to have been used by Polish prisoners-of-war (Uznanski, 1957, p.3).

The "town" consisted of about ten permanent log buildings and approximately 50 huts. The huts were square or circular, perhaps 20 feet in diameter (Fig. 10). Some were on the surface and others were several feet below ground level. It was not evident whether the huts had sunk because of thawing of the permafrost or had been dug out for better insulation. All huts were surrounded with sod or double walled, two-feet thick, and stuffed with dried grasses. A long fiber called paklya was used for caulking between logs in the walls. Rugs were hung on the walls to keep the cold out. The town was situated on a flood plain in an area dotted with many residual lakes connected by meandering streams. The country was barren of trees and there was little grass. Pigs were being raised. Many large husky-like dogs made their presence known.

It was announced that the planes were grounded because of bad weather at the next stop, Yakutsk on the Lena River. Attempts to see the weather reports were unsuccessful even though we tried to explain that there were several qualified meteorologists in the group who could interpret them.

The main building housed the radio shack judging by the aerials leading to tall poles around the building. Two thermoscreens of the type used to house meteorological instruments were visible, but no instruments were seen except for a swinging plate that presumably was calibrated to measure the wind force by the angle of deflection. One pilot remarked jokingly that it would take a hurricane to move the steel plate. We all began to understand why the data being received from such stations were not compatible with that from the more sophisticated instruments in the U.S. Needless to say, we did not get inside the main building, and it appeared that no decision was to be reached until direct orders were obtained by higher Soviet authority.

It was still light at 10 P.M., being just below the Arctic Circle. The electric lights available were turned on after midnight for the few hours of darkness that remained. Dinner was to be served, so there was a rush to freshen up. The outhouse was communal for both men and women and the choice of a hole or trough was used with some finesse and no apparent embarrassment. The anteroom to the dining hall had a can of heated water on the wall that operated like a balm dispenser. When the plunger at the bottom was pushed up a handful of warm water was released for washing hands and face. The cloth towels on the rack were for community use.

Our first dinner in Siberia was unusual to say the least. Imagine having salmon caviar (fish roe), cheese, black and white bread and fruit (small oranges and yellow apples) as appetizers. The black bread was delicious in taste, very heavy in body, and moist—almost a meal in itself. Disaster followed as toast after toast was made to honor the leaders of each country. Even though there were toasts to one's health (na zdorovia) it had just the opposite effect, especially if followed by the phrase "bottoms up" (pai do dna). The few who were still able to sit at the table had a large portion of borsch (a beetroot soup) filled with carrots, cabbage, and potatoes. The last course consisted of pork steak and fried potatoes. It was one of the rare times we saw a whole piece of meat; meat was usually diced, shredded or ground. Plenty of time was given to enjoy each course, but many of the diners were just too sick from the "spirits" to partake.

After dinner those who were able watched the men of the town play a game called "gorodkee." It resembled bowling in a way: Five round plugs 1″ × 6″ were set on end on a board about 5 feet square. One stood about 50 feet away and attempted to knock off the standing plugs with a wooden club, 1½″ in diameter and 4 feet long, thrown side-arm fashion. Each plug knocked off gave the thrower an advantage by moving a fixed distance closer. For each "inning" the plugs were set in a different pattern at the choice of the opposition. While watching the game an attempt was made to learn a few Russian words from the very congenial players who had fun learning a little English. The Russian phrase book (TM 30–644) provided by the War Department was limited to more practical expressions such as "Don't shoot," "Where am I," and "Where is the toilet."

Most of the group slept on top of the cargo in the planes. The unlucky commanding officer was given the honor of having a bed in one of the buildings—he sat up most of the remainder of the "night" due to bed bugs. Others constructed raised platforms in the dining hall by putting planks, which had been stored outside, across logs. Unfortunately the frozen planks thawed from the body heat and the sleeping bags became wet. Next morning, few could face the breakfast of fish cakes in sour cream. Clearance had been received during the night, but another day had been lost when passing the international dateline that runs through the Bering Strait.

At 7:30 A.M. a tired and sick group manned their planes. The first plane attempted to take off, but after rolling down the runway it was discovered that the blocks used to hold one of the ailerons and the tail rudder in place while parked had not been removed. It was evident neither the flight crew nor the ground crew were in the best condition to inspect the aircraft. The result was a sudden stop at the end of the runway. The cargo contained the segments of the metal antenna towers; as Lt (jg) Uskievich recalled, the crew and passengers barely avoided being skewered as the frames lurched forward into the cockpit . After retrieving the blocks, the second attempt at takeoff was more successful and the flight was on its way to Yakutsk.

The long flight leg.

Aside from the early morning haze, the weather was excellent all the way to Yakutsk. The tundra (treeless grassland) was still relatively green and the multi-colored growth around the multitude of lakes and ponds was spectacular. The growth rings of specific varieties of plants and shrubs were marked by sharp color changes. It became evident that some lakes had been completely overgrown by a spongy mat. Ground travel over such country would have had to take place during the winter months. In other regions there were textbook examples of the network of polygonal arrays of rocks accumulated by the heaving and thawing of the permafrost.

The flight was proceeding at a relatively reasonable altitude above the terrain on the advice of the Soviet navigator until some peaks of the Kolyma Range appeared, and the pilot immediately pulled up to 10,000 feet! The plane passed over the town of Oymyakon, which then shared the record with Verkhoyansk as the coldest inhabited place on earth, $-90.4°$ F.[25] (Temperatures some 38.2° lower, that is, $-128.6°$ F., have since been recorded by the Russians on 21 July 1983 at the Vostok station, elevation 3420 m, in Antarctica, near the south geomagnetic pole.) It was alleged one could literally throw a pan of water up into the air and have it return a disk of ice! As we were to experience later, even Russian vodka freezes around $-40°$ F.

Farms were seen within several hundred miles from Ya-

---

[25] It should be noted that a mercury thermometer freezes at about $-38°F$, hence other methods are required to record lower temperatures. Other liquids such as ethyl alcohol, toluene or pentane have been used in glass tubes to record temperatures down to $-200°F$. In radiosondes, a package of instruments sent aloft on a balloon, the temperature is measured with a crude bimetallic strip. Other devices employ the resistance change of platinum or copper. For example, a temperature of $-120.3°F$. $(-84.6° C.)$ was recorded with a strain-free, platinum-resistance thermometer (calibrated at the dry ice point) at Dome C station (74.65° S, 123.17° E), Antarctica, elevation 3240 meters, by Dr. Charles Stern, Department of Atmospheric and Oceanographic Sciences, University of Wisconsin, at 0900, 26 August 1982 (personal communication, 11 July 1995). It is a difficult measurement even when great care is taken to isolate the instrument from the direct sun and louvers are provided for free flow of air currents, for example, as in a thermoscreen (Fig. 40).

kutsk. Most farms were on dried lake beds or the soil was tilled around the many lakes. One surmised from the hay stacks that grains were grown. The haystacks were raised on log rafts so they could be towed across the lakes or floated down streams to the collective farm centers. Harvesting was done by hand with a two-handled scythe.

Yakutsk—more delay.

The same ploy was used: The weather at the next stop, Khabarovsk, was bad and the flights grounded. In short, more instructions were needed from Moscow. Although the local time was 9:30 A.M., lunch was served to the group. It consisted of a red cabbage and carrot salad, another salad of white cabbage and green cucumbers in vinegar, goat meat hamburgers with mashed potatoes, hot chocolate, and a pastry dessert (kvorost) akin to our sugared donut.

The airfield was on a river flat about five miles from the city proper. There was a single air strip and no hangers. About 30 DC-3s and 10 B-25s were on the field in addition to approximately 20 old biplanes, apparently operational, parked on one side of the field. Women gassed the planes from steel gas drums and a hand pump, even though Yakutsk was a major terminal for the Russo-Japanese front. The DC-3s took off without warm-up in an effort to save fuel.

The city of Yakutsk lies on the Lena River, a major north-flowing waterway utilized for hauling coal, lumber, and grain. The river was braided, about a quarter of a mile wide, with large sand bars. A refinery was near the city, but the roads seemed to terminate about 20 miles out. Some mining[26] was visible, but little did we realize that the famous dia-

---

[26] Mining and drilling in permafrost requires a special approach because of the freezing conditions. For example, water used to remove the particles generated during drilling has to be saturated with $CaCl_2$ to lower the freezing temperature. The drill stem will twist off if circulation stops and the water freezes. Alternatively, air drilling can be used, but the particles produced are usually too fine to identify the formations encountered. In addition, it is necessary to add $CaCl_2$ to the cement in casing a hole to promote setting. In open pit mining, the principal danger is from slope failure as the permafrost thaws. As much as 30 percent more explosive is needed to rubblize frozen material in contrast to unfrozen material. Dust generated by mining cannot be controlled by the usual way with a water spray. A hazard encountered in underground mining is the presence of saline pockets

mond-bearing volcanic pipes were in the Yakutia region.[27] Almost all the buildings were of unpainted rough timber. The roofs of the houses were not shingled, but covered with long overlapping boards. The "factories" were housed in large two-story buildings, and could only be distinguished from living quarters by tall metal smoke stacks. The "factories" served more as warehouses for the parts and materials that the women took home to carry out a specific operation by hand. In this way, the women could maintain a home for the children as well as contribute to the war effort by preparing materials for assembly by others in the "factories." The windows were doubled, containing a miniature window (fortochka) for ventilation, and the frames were scalloped in various designs and rarely painted (Fig. 11). Little light got into the rooms because of the large number of plants placed on shelves inside the windows. Many of the houses were constructed of 4" × 4" logs, pegged together and caulked with moss. Some of the houses were grouped to form an enclosure for a community barnyard for cattle. Water was carried from wells in two buckets hanging from each end of a pole balanced over a shoulder, Chinese style.

The outdoor bazaar was primarily a vegetable market: onions, beets, turnips, cabbage, tomatoes, potatoes, carrots and a berry similar to a cranberry. Large round loaves of bread were offered for sale. The shopkeepers used an abacus to add up the bill. Stands along the street sold "fruit water." A drugstore, restaurant, and several hotels were identified, but the numbering system was baffling. The hotels, for example, were numbered not consecutively on the street, but

within the permafrost zone. The Russians have had considerable experience in permafrost mining, but will quickly admit to the need for better techniques.

[27] The kimberlite pipes in the Yakutia region noted for their large and high quality diamonds are actually near the Arctic Circle, 600 miles to the northwest of Yakutsk. There is a spectacular pile at least a foot high of single crystals of diamond in the Kremlin Museum that are larger than the end of a thumb! On a visit some years after the MOKO expedition, I showed too much interest, not in the diamonds, but more importantly, the adjoining smaller pile of zircons that occur with the diamonds and could be used for age determination, and was promptly escorted out of the Museum. Eventually samples were obtained through the cooperation of a member of the Soviet Academy of Sciences, Prof. V. S. Sobolev, and the age of the zircons was found to range from 148 to 443 million years depending on the pipe locality.

Fig. 11. A typical old style house. Photographed in Shimanovska, 400 miles northwest of Khabarovsk, by Mr. Raymond A. Rzeszut in May 1996. The preservation of these unique houses, some well over 100 years old, was not a high priority in the Soviet Union. The wood plank roofs have been replaced with corrugated sheet metal.

by the order in which they were erected in the city, e.g., Hotel #1, Hotel #2, etc., as the need was determined by the government. Each hotel had about 40 rooms, with several guests to a room. Men and women shared a community wash stand, but the toilets were separate and without toilet paper. No bathing facilities were evident. There was an electric light in each room; the light was draped with a fancy shawl over a wire frame in order to cut down the glare. The cast iron beds, without springs, were fitted with two sheets, pillow, and a thick wool blanket. One phone served the hotel. All gathering places had a radio speaker over which government announcements could be made. It could be turned down, but never off! Between messages, choral singing and symphonic music was broadcast. Billboards along the sidewalks provided the details regarding government announcements as well as news.

Board sidewalks lined the very broad streets—at least four lanes wide. The spacious avenues were a necessity as fire breaks. In the winter, accessibility to water was limited, so when one house burned, the entire block went up in flames.

Only two "paved" streets were seen. The paving consisted of 6″ logs set vertically and the interstices filled with dirt. The logs were subject to heave and thaw, and were pounded level on a regular basis. The permafrost (permanently frozen ground) actually extends down to at least 450 feet, but only the top 8 feet thaws out during the summer at Yakutsk.

Again most of the automotive equipment was of U.S. design, but not necessarily built in the U.S. In addition to Studebaker trucks, a large number of relatively new Model A Fords were visible! Henry Ford had sold the machinery for making Model As to the Soviets on 31 May 1929, and they were produced in Nizhni Novgorod, now called Gorky, 250 miles east of Moscow.[28] The main transportation, however, was by horse-drawn carts. There was a large yoke over the horse's heads that seemed excessively heavy for horses that were about a hand shorter than U.S. riding horses.

There was a more friendly atmosphere in Yakutsk than in Velkal and a genuine desire to make the brief layover comfortable. Some of the men were quartered in hotels and most of the officers were housed at the airfield in Soviet Officers Quarters (see Fig. 12 for unique dining service from the officer's mess). Some meals were obtained at a local restaurant. Supper consisted of borsch, hamburger, french-fried potatoes, kumquats in heavy syrup, bread made with sour cream, and beer. The entire group was taken to the theater that held about 200 people. As we learned later, people already in their seats were asked to leave to accommodate the group! A 15-piece orchestra played and beautiful choral singing came from behind the curtain before each of the three acts. The stage settings were exceptionally well done. Curved drop cur-

---

[28] The Ford Motor Company signed a contract on 31 May 1929 with the Supreme Council of National Economy of the USSR and the Amtog Trading Corporation for the "exclusive license to manufacture, use and sell in the USSR, as now constituted, under all present patents and future patents and inventions owned or controlled by the Company or its subsidiaries covering all materials and component parts and units of the Company's Model A cars and Model AA trucks." The Ford Motor Company agreed to supervise the building of the plants for an annual production of 100,000 cars. The first car came off the assembly line in February 1930 and the first truck was produced in January 1932. The contract was canceled in the spring of 1935 as the world-wide depression set in. Although the cars and trucks were especially suited to the road conditions in the USSR, the major contribution was the training of Soviet engineers in the U.S. and establishing the assembly line technique in the USSR (Wilkins and Hill, 1964, Chap. 10).

Fig. 12. Dining table accessories used at the Red Army Mess in Yakutsk, Siberia. From left to right; rock salt dish, pepper shaker, and silver glass holder (podstakannik) for hot tea. In front of the glass holder is a silver bar on which to rest soiled utensils when not eating. The Russian letters "PKKA" stand for "Workers' and Peasants' Red Army."

tains in the corners at the back of the stage housed the scene props. It was a very efficient way to set the scene on one side while acting proceeded on the other side of the main stage. The play, in brief, was about the attempts of a soldier returning from the front to locate a girl who had saved his life. He looked up all the Russian equivalents of "Mary Smith" in town and gets into predicaments with each. Even without understanding the language, the play was very humorous. Oh yes, in the end he did find the right "Mary Smith."

The next morning the officers witnessed a scene within sight of their quarters that was in such great contrast with the events of the night before that it cannot be forgotten. Hundreds of political prisoners, so identified by the Soviets, were working on the road in a simplistic and primitive way. Rows of prisoners lined across the road shuffled along moving 6' long pikes up and down to loosen the irregular chunks of frozen mud that roughened the road. Behind them came others who scooped the chunks of dirt into piles with their bare hands. Still others followed in pairs carrying a wide board on which the dirt was piled. The dirt was carried back to fill holes or across the 6' deep drainage ditches that bordered the road. Discipline was perfect and for good reason: those who worked got more food! There appeared to be no division of labor between the men and women. As experience was to show, there was no task a Russian man could do that a Russian woman could not do as well.

The planes took off at 6:30 A.M. after being serviced, but no briefing was obtained. The weather always seemed to improve suddenly when clearance arrived from Moscow.

Khabarovsk—at last.

After a five-hour flight in average weather, the planes landed at 11:30 A.M. on 26 August 1945. The pilots of Squadron VR-5 had done a superb job flying a route over which U. S. pilots had never flown before. Soviet Major General Yarmilov[29] met the planes that had been parked near an access road at an isolated spot at one side of the field away from the terminal facilities. The field was about 5 miles southeast of the city and consisted of two parallel runways oriented NE-SW. The runways were laid with 6' wide hexagonal concrete slabs to form a 5,000' long track. Seven hangers were seen in the distance, and all the automotive equipment was U.S. made. Revetments, earthen mounds constructed to provide protection from bomb shrapnel, were located on the periphery of the grass area, and the taxi roads were constructed so that fighters could take off directly from the opening in the revetment.

A three-story tower housed the control room and radio facilities. Directions were given to planes, but traffic "control" was a misnomer. A flight of over 100 bombers and fighters returned from Manchuria. It was not evident whether the planes were returning from a raid or were no longer needed in the Manchuria theater. After a flare was fired it seemed that every pilot had an equal opportunity to land. The result was similar to the dispersal of ducks being fired at from all directions. Planes landed in any direction irrespective of the wind and anywhere there was an open place on the grass. When trouble occurred a flare went up to wave off pilots. The reason for the apparent lack of rules on landing was later explained by the fact that radios had been removed

---

[29] Relationships with Soviet officers were maintained on a formal basis, and their first names were rarely if ever used even on signed documents. For that reason only rank and family names are known to the writer and other Unit officers. My Russian colleagues indicate that the custom is common today, perhaps modified with the initial of the first name only.

from all the planes except the flight leaders and given to tank commanders. Planes were expected to fly contact; that is, navigation was to be done by direct observation of landmarks and visual reference to the flight leader! In addition to the U.S. aircraft of P-40s, A-20s, and C-47s, some Soviet TUP-2s and YAK-9s were spotted, and among them some rare Japanese transports and German Messerschmitts. In all, there might have been 500 airplanes on the ground before we left the field. As the taxiing planes came by to see the comparatively gigantic U.S. Navy R5D-3s, the canopies were slid open, and smiling, waving women appeared in view. One got the impression that the entire flight was manned by women crews. Women aviators had served with great distinction as front-line combat pilots against the Germans. There were at least three air regiments composed entirely of women, including mechanics, armament technicians, radio operators, gunners, navigators, and pilots of fighters and bombers (Myles, 1981).

An incident occurred during the unloading of the U.S. Navy planes that revealed an important aspect of the Russian character. On board was a 3-ton diesel-powered electrical generator for running the radio transmitting equipment. The largest crane at the airfield could only support 2 tons. Nevertheless, the crane was hooked up to the generator and on the first attempt to lift, the tires blew out. It was suggested that a platform and ramp be built of wood so that the generator could be rolled into a truck. On orders that the generator was to be unloaded at once, a Soviet colonel put his fist to his chest and said "Ruskie dyello," that is, it *will* be done by the Russians. Some 40 to 50 teenagers, all very strong looking, from a nearby Young Comrades Camp were assembled, close packed, at the rear door of the plane. The generator was rolled out and eager hands grabbed the skid until the group bore the entire weight of the generator. Knees buckled but only from having the good sense to use one's legs for lifting power. The first impression was that they would all be crushed, but they held and walked lockstep to the waiting truck. What a performance! From then on, whenever we heard the phrase "Ruskie dyello," we knew the job would be done no matter what the cost. Needless to say, our respect for these hardy people skyrocketed.

3. INSTALLATION OF MOKO BASE

Summer camp.

ALL THE GEAR was quickly loaded into Studebaker trucks, and personnel, including the flight crews, were told to stand in a canvas covered truck. The personnel truck took off at great speed down the dusty road, and after a few miles stopped in a forested area. It only took a few minutes before the idea crossed our mind that we were about to be eliminated at this isolated area.[30] The lack of official enthusiasm for our presence and general suspicion about the intent of the mission was felt by all. Fortunately, the remainder of the convoy eventually caught up and our state of mind suddenly changed—temporarily. We proceeded another twenty miles east of the main city to a compound that was to serve as the MOKO base. It was a small summer camp for rest and recuperation of factory workers. (The camp kitchen kettle had a posted capacity of 160 liters, about 170 quarts, which would indicate that over 150 people might have attended the summer camp.) A new seven-strand barbed wire fence completely surrounded the camp and a guard holding a submachine gun was positioned at the main gate. A fence can be interpreted two ways, depending on which side you are standing. One view is that the property on the inside is being protected. On the other hand, those on the inside could not get out. Somehow the latter version was more in keeping with the prevailing attitude. The commanding officer was taken on a brief

---

[30] No one in the group was apparently aware of the Katyn Forest massacre of the more than 4,443 Polish officers and a few enlisted men and civilians by the Soviets in the spring of 1940 (Zawodny, 1962). At any rate, no one suggested under the circumstances that evasive action be taken by the group. About 15,000 Poles in total were missing as Prisoners of War (POWs), but it is alleged that a large number were sent to Siberia not as POWs but as sentenced forced laborers, thereby losing whatever protection they might have received through the Geneva Conventions. The change of status was decided by the Head of the NKVD and the POWs were not so informed (personal communication, Prof. J.J. Danielski, 7 February 1996). From the Soviet viewpoint there were no Polish POWs in Siberia. No Polish prisoners either as POWs or as forced laborers were seen by members of the MOKO Expedition. Perhaps it was fortunate those chilling events were not known to the group and drastic action avoided.

Fig. 13. Recreation Hall for enlisted men.

tour of the physical plant by the Soviet General and decisions
were made on what was to go where. The space appeared to
be adequate for carrying out the mission. Soviet labor unloaded the trucks and personal gear stowed. In typical Navy
style, there was a separate officers mess and quarters, and
enlisted men had their mess and recreation hall (Fig. 13), but
their quarters consisted of a renovated, many-stabled horse
barn (Fig. 14). Local women from the nearby village of Knaz
Volkonka were assigned cooking duties (Fig. 15) and cleaning services.

The flight crews returned to the field and all planes
took off at 10:00 A.M. for the return trip to the U.S. The
needs of all the departments were drawn up in the way of
office equipment. The Soviet liaison engineer met with the
commanding officer and he accepted all requests without ex-
ception. The Russian phrase "Budyet zaftra," it will be done
tomorrow, was heard often, but we learned that the real
meaning was closer to the Spanish word "mañana," that is, it
will never be done! In the meantime all hands turned to
erecting three 90-foot antenna towers. The installation of the
radio receivers was given number one priority. A temporary
antenna was raised by running a wire to a 25-foot pole in

Fig. 14. Quarters for enlisted men, a renovated horse barn.

Fig. 15. Maria A. Zorina, maid for Capt. Cumberledge's dacha, believed to be an NKVD agent.

the Soviet garrison across the bordering ravine. The pole was made of scrap pipe welded together in a local shop. After a day and a half, the guy wires were in place and the three towers erected (one can be seen in figure 16). The erection crew, who arrived early in the morning, were unnerved by the sound of rifle fire. In answer to their expressed concern, the jeep driver from the Soviet garrison said, "not to worry, it was just the morning firing squad"! Mounting an anemometer and weather vane at the top of one of the towers proved

Fig. 16. One of three antenna towers erected for transmissions to Guam. Note anemometer and wind vane on top for recording wind speed and direction.

to be a rather harrowing experience. Because of the lack of volunteers, Lt. Starke, the senior weather watch officer, made the climb and successfully installed the instruments.

Negotiations.

The Commanding Officer and interpreters met at 9:30 A.M. 28 August 1945 with Soviet Major Vishvarka, the liaison engineer (Fig. 17), concerning improvements and winterization of the summer camp provided by them. Next, the CO met with Captain (2nd rank) Tikhonovetski of the Soviet Navy, commanding officer of the Khabarovsk Weather Office. Initial introductions were made as to the purpose of the MOKO Unit to the captain. A free exchange of information was offered by both our CO and the Soviet captain. An appointment was made for 30 August 1945 at which time the Soviet captain promised to supply detailed information as to the location of Soviet weather stations east of 110°E and outline their reporting system. (The group in Petropavlovsk was to collate the weather data from the Soviet stations east of 140° E.) This session was followed by an evening meeting with Major General Yarmilov at the CO's quarters on 29 Au-

Fig. 17. Main building of the Summer Camp used for offices and officer quarters. Soviet major Vishvarka, an engineer, serving as Liaison Officer. Note that a US flag is flying on a separate pole at the station along with the USSR flag.

gust 1945. Again the TOP SECRET nature and importance of the mission was emphasized even though the war was presumed to be terminated. The general concurred. He agreed that a Soviet liaison weather officer work with the MOKO Unit in place of an engineer. The general also stated that military air transportation was available to the CO for travel to Vladivostok or Moscow.

The next day, the planned meeting with Captain Tikhonovetski took place and the chart showing the location of the Soviet weather stations east of 110°E was provided.[31] The broadcast schedules of weather stations in only the northeastern part of the Soviet Union were also provided with the promise that others would be forthcoming. Specifically discussed was the weather report collecting system at Khabarovsk and the means of sending those reports to the MOKO Unit, but the CO was told that those issues had to be reviewed with Major General Yarmilov. Thus, it appeared that

---

[31] The author does not recall any of the details of the chart, but there were probably no more than a dozen stations reporting. The area east of 110°E in Russia is now served by about 65 stations according to the data entries shown on northern hemisphere maps prepared by the U.S. Weather Service.

Fig. 18. Radiomen at radio receivers.

some headway had been made—at least in the gathering of
pertinent weather information in eastern Siberia—in addition to the collection of local observations at the MOKO base.
Because the data from the Soviet stations were essential for
the formulation of an accurate weather forecast, one needed
to know not only the exact location of the reporting station
but also the elevation, topographic features, and proximity
to bodies of water. Although the aerologists of the MOKO
Unit were experienced in ship-board, single-station methods
of forecasting, having a large body of data from a wide range
of stations was a distinct advantage. The existence of the Sik-
hote Alin Mountain range between the MOKO Base and the
Japanese Islands was a major factor to be considered in any
forecast, so accurate knowledge of the reporting stations
was essential.

Meanwhile radio reception was established at 8 P.M. on
29 August 1945 (Fig.18) and work on the transmitting equipment began. More personnel arrived with the third and presumed final flight on 6 September 1945. There were 19
officers and 41 men now at the MOKO base (Fig. 19). Their
names and specialties are listed in Table 1. (Specific names
have not been used in the narrative up to this point because
of poor recollection on which flights the various officers and
men had arrived). A photograph, taken by a Soviet photographer of the commanding officer, Capt. A. A. Cumberledge,

Fig. 19. Officers and men of the MOKO Unit. Photograph taken on 22 November 1945 by Soviet photographer. The officers in light colored pants in the first row are from the U.S. Army. The Soviet officer in the long coat in the first row, 5th from the right, is Major Vishvarka.

Fig. 20. Commanding Officer Capt. A. A. Cumberledge (left), Soviet Major Vishvarka, and Executive Officer Cmdr. E. E. Butow (right).

and the executive officer, Cmdr. E. E. Butow, is given in Fig. 20. Although a formal peace agreement with Japan had been signed on 2 September 1945, it was still not clear how the Japanese military would react. The abortive coup on 14–15 August was evidence that all military units were not willing to accept surrender. The establishment of the weather base was undertaken, therefore, with a full understanding of the necessity and urgency of the mission.

A conference with Lt. Vederakin, General Yarmilov's adjutant, revealed some progress and, conversely, some startling constraints. A direct phone line had been installed between the MOKO base and Khabarovsk, but was not connected yet. Identification documents were requested for the officers and men of the unit when they left the base. In addition, a list of available Russian foods was requested so that a diet closer to American "standards" could be prepared. For example, the kasha, an oatmeal-like porridge, was shifted from dinner to breakfast. The small, excellent, Russian pancakes (blinchiki), usually served as a dinner appetizer, were also shifted to breakfast. The bowls of fish eyes, a highly prized delicacy for the Russians, was deleted, even though it was pointed out that a seagull eats them first! The substitute of herring with an onion ring around its head was not acceptable either. Several days later a list was received by the CO showing the food ration for one person per day. Food ration cards were to be furnished so that the unit could obtain food at its own expense in Khabarovsk even though it was understood the Soviets were to feed the units. One officer kept a gallon can of raisins on his desk as a supplement to his rations. When the second flight returned to the U.S., someone had slipped a note to the pilot for delivery to Lt(jg) Nelson,[32] an aerologist who was acting temporarily as supply officer for the unit remaining back in Seattle, to send all the food they could find room for! Nelson managed to round up provisions for 60 men for 90 days with the help of an understanding warrant supply officer at Pier 91. It did not take long to realize that the Soviet diet and paucity of food was going to be a major problem. Eventually it was necessary to get two railroad car loads of survivor's food from Murmansk.

The adjutant also noted that only Soviet drivers were permitted to operate the American Lend-Lease trucks and jeeps provided to the unit. It was difficult to persuade the drivers that the vehicles were indeed made in the U.S.A. because all the instruments were labeled in Russian. The name "jeep" comes from the sound of the first letter of the first two words in the phrase General Purpose Vehicle, sounds that are readily reproduced with three Russian characters. (A more hilarious argument arose over the origin of the alleged "famous Soviet Douglas airplane"!) Although it was claimed Soviet drivers were noted for their regard for safety, that assertion was difficult to believe after taking an "ice spin." The trick was to drive the jeep off the low cliff bordering the Sita River at high speed, set the brakes in the air, so that the jeep

---

[32] Lt(jg) Frederick A. Nelson had served on the USS Hornet (CV 8) in the first year of the war under then Lt. Cmdr. Cumberledge, aerological officer. He had also served with Lt(jg) R. W. Monical in the weather office at the Naval Academy before WW II. The author was fortunate to serve with Nelson at the Fleet Weather Central in Seattle during April–June, 1943, while awaiting the fitting out of the USS Baffins (ACV-35, Auxiliary Aircraft Carrier, later decommissioned for service in the British navy). During that period I flew as aerologist on the nonstop transport runs that had just been initiated between Seattle, WA, to Kodiak, AK. The uncertainty of the weather in that region was a challenge to the most experienced meteorologist, but provided excellent training for the forthcoming MOKO expedition.

would spin on the ice on landing. Apparently that fun trick was not covered in the safety manual, but it certainly relieved any accumulated boredom from the cold weather. The drivers also took great delight in coming as close as possible to persons walking along the road without taking the buttons off their clothes.

The question of fraternization with women was addressed directly by Lt. Vederakin. Fraternization of Soviet citizens with any foreigners, including Americans, was not permitted. He modified this statement slightly by saying that only certain women would be permitted to associate with MOKO personnel. The reason given was that venereal diseases were prevalent, but it became evident that "certain women" included the "cleaning women," who were there to observe and not just clean. One officer believed that at least one cleaning woman (Fig. 21) was too obvious in her observations to leave any doubt as to her primary mission. As will be seen later, other women encountered "casually" through apparently random, unplanned events had similar assignments.

Communication with the Soviets as well as the U.S. Navy in Guam was a perpetual problem. None of the Soviet high command would commit himself to a commercial link between Moscow and the MOKO base because it had not been discussed at Potsdam. Major Vishvarko said that Red Army telephone facilities were available to Moscow. No coded messages could be sent from the Khabarovsk Central Telegraph Station. The transmission frequencies of the MOKO Unit were supplied to Chief of Communications in Khabarovsk, but approval had not been forthcoming. The Soviets did, however, grant permission for a teletype circuit between Khabarovsk and the MOKO base. Initially, the commercial telegraph was utilized. Lt(jg) Uskievich made daily trips to Khabarovsk to pick up messages addressed to the MOKO Unit. The four-hour round trip was made by jeep no matter what the weather or the condition of the roads. Lack of progress in communication of Soviet weather data was due in part to the absence of Captain Tikhonovetski from the Soviet Hydrometeorological Office in Khabarovsk. His assistant Lt. Pede did not know any of the details and was unable to take any action while his superior was gone. We soon got used to

Fig. 21. Alleged cleaning woman, assumed to be Soviet agent.

the technique employed to delay decisions: The senior officer is in a meeting, out of town, or is awaiting instructions from Moscow. (The Soviet technique seems to have also been invented simultaneously in Washington, but it is still called "red tape," originally after the red tape used to tie official documents).

The third flight of planes arrived on 6 September and was held on the ground because of bad weather. By the 10th of September the weather cleared sufficiently along the route home for the planes to take off. The CO had accumulated a long list of equipment needs and decided that one officer should return to the U.S.A. to acquire missing transmitter parts, a spare electrical generator, and, most importantly, to find out why coded messages to the Unit from Guam could not be broken with the code books in hand. The author was tagged for the return flight probably because the weather forecasting functions were not yet operational.

Fourth Flight Hassle.

The return flight of about 26 hours was made via Yakutsk, Markova, Fairbanks, Anchorage to Seattle. Markova was somewhat of an improvement over Velkal in that the runway, equipped with lights, consisted of interlocked, perforated, metal planks[33] made by the U.S. and used as landing mats in the fast construction of temporary airfields. There were about 100 dwellings. The windows contained small pieces of glass that were overlapped to avoid waste. The coat hooks were simply wooden pegs. Caribou hide covered some of the doors, and most of the villagers were dressed like Eskimos. The only entertainment in sight was a hand-wound Victrola well stocked with Columbia and Victor records. A volleyball court and a small bandstand provided other outlets. As in other villages on permafrost, the roads were high crowned with deep drainage ditches. At Markova they had the luxury of small curved wooden foot bridges over the ditches.

In Seattle it was learned that the reason our coded messages could not be deciphered was that the codes had been changed at the end of August! The generator, transmitter parts, and supplies were rapidly acquired and loaded. One item, an army theodolite (an instrument used for measuring horizontal and vertical angles and heights of a balloon) requested by U. S. Army Major Bristor, turned out to be particularly troublesome. Even though Lt(jg) Nelson was not persuaded that it was superior to the Navy model, he located one in Spokane, WA. Air transport for Nelson was arranged, but he discovered that the instrument was signed out to a supply sergeant who obeyed the rules. The sergeant's commanding officer was on leave in California, so Nelson called his home only to find out that he was at a movie. The officer was paged at the movie, the situation explained, and he called the sergeant giving him permission to release the theodolite immediately. Nelson made it back from Spokane just in time for the departure of the fourth flight.

---

[33] The pierced-steel-plates (PSP) were developed by the U. S. Army Waterways Experiment Station at Vicksburg, MI, because of their expertise in soil mechanics. The landing mats were first tested in November 1941, during training maneuvers near Marston, NC, by the 21st Aviation Engineers Regiment, and became known as Marston mats (M. C. Robinson in B. W. Fowle, 1992). Each interlocking plank was 15 inches wide, 10 feet long, and weighed 70 pounds. An airstrip 100 feet wide and 2,500 feet long could be prepared and assembled in as little as four days. Some 2 million tons of mat were produced during World War II.

The MOKO Unit had been reclassified as RE-STRICTED so there were fewer constraints on talking about the needs of the Unit to meet its mission. Washington even said plans were underway to send a plane a month with supplies and to rotate personnel. The first plane of the fourth flight got off at 9:30 A.M. on 16 September 1945 and landed in Fairbanks after a stopover in Anchorage. After all his trouble, Nelson was rewarded with a flight on a plane with a broken heating system—good preparation for what was to come.

The fourth flight was not permitted to take off from Fairbanks on the orders of Lt. Colonel Khallinnokov of the Soviet Detachment. The original agreement was for three flights of three planes each and that agreement was finished. Once again, the routine began about getting permission from Moscow. This time it took six days before a decision could be reached! The fact that almost 8,000 Lend-Lease airplanes had been delivered via the Siberian route (Cohen, 1988, p. 34) did not seem to influence the decision. Next, the delay was the weather. It was so bad that reconnaissance flights taxied directly into the hangers, with the doors opened and closed with minimal clearance, for regular servicing and discharge of crew. Snow tunnels were available between the hangers, quarters, and mess hall, so one did not have to endure the piercing wind and cold. Because of the strong temperature inversion aloft in the Fairbanks basin, the heavy flying suits were too warm several thousand feet up and had to be shed. The inversion was just one of the many anomalies of Arctic weather. Eventually the fourth flight managed to reach Nome.

On 26 September 1945 clearance was obtained but trouble struck again. An engine on the plane carrying the electrical generator had to be feathered just after takeoff from Nome. With considerable skill the pilot did a 180° turn and managed to hit the end of the runway moving down wind. The plane skidded off the far end of the runway and damaged the nose wheel. Fortunately the 3-ton generator stayed in place. It was also fortunate that the carboys of battery acid on which the author was riding did not break! For reasons that were never clear, the CO of the MOKO Unit had been informed by the Soviets *two days before* that the fourth flight had been forced down at a place called Achita. The origin of this erroneous report could not be determined.[34]

While the plane was being repaired I had an opportunity to visit the permafrost station on the outskirts of Nome. A hole had been drilled to monitor the changes in temperature with depth. The maximum depth of thaw was only 2½ feet at Nome. Building roads on tundra is the major problem (see, for example, Ferrians, et al., 1969). If any sort of road cover is laid, it acts like a blanket and the ground thaws. Ground water trapped between a new surface freeze and the permanently frozen ground moves to the road and literally pushes it up. Small ice glaciers are observed along the road. The same problem is encountered in laying a foundation for a house or building. It is not uncommon for water to burst the blanketing foundation and fill the house with water! The Russians have tried to solve the problem by sinking piles deep into the permanently frozen ground with a sleeve in the area of potential thaw. By the strange cocked angles at which some of the Nome buildings sat, it was evident that a solution to the permafrost problem had not been achieved, even with concrete rafts and 12-foot deep pilings.

The town of Nome consisted of about 300 houses on old beach ridges. Several abandoned gold dredges that betrayed its earlier days of glory were seen in the ship canal. Freighters laid off the coast and supplies were lightered off on to barges because of the shallow pebble beach. The population appeared to be dominated by Eskimos who wore parkas of reindeer hide with the fur inside. The teeth of the Eskimo women were worn down to the gum line from chewing on the walrus hides to make them pliable. It seemed incongruous for the Eskimos to have an outboard motor on their walrus-hide boats! The Eskimos from the nearby King Island in the Bering Straits brought in carved ivory for sale. Walrus tusks and teeth were readily available as well as pieces of fossil mammoth (mastodon) ivory.

By September 27 the fourth flight was assembled at Markova for the next leg to Yakutsk. That night an incident

---

[34] The town of "Achita" could not be identified on either Soviet or U.S. maps of eastern Siberia. Alternative spellings such as Akita, Chita, or Achinsk were found in unlikely areas.

took place that almost terminated the flight. One man had brought a large quantity of chocolate bars and stored them for safe keeping in a chest of drawers. During the night a very large rodent—almost the size of a ground hog—gnawed a hole through the wooden chest and began to eat the chocolate. A pilot awakened by the noise, drew his 45-caliber pistol, and blew the rodent and chocolate bars to smithereens. Bedlam ensued as the guards charged the room. Fortunately, they saw the humorous side of the chocolate catastrophe and sleep was resumed. The flight was on its way to Yakutsk the next day.

The usual delays in getting permission to take off from Yakutsk for Khabarovsk were experienced and a new delaying tactic was imposed. Soviet mechanics removed a distributor from one engine of each plane on the grounds that they were faulty and needed to be tested. After two days of that nonsense, the pilots took the parts from the engines of one airplane and outfitted the other two. Without interference—but not without some risk—the two planes took off for Khabarovsk at 7:20 A.M. and landed safely there about four and a half hours later. The planes were not buzzed nor fired on by Soviet aircraft during the flight. The third plane containing the standby electrical generator was retrieved on 3 October by transferring parts again from the returning planes. On landing at Khabarovsk the problem of unloading the three-ton generator had to be faced again. This time good sense prevailed, and a ramp was constructed from the plane to the truck. The ramp, however, was too steep, and it was not possible for the assigned young Soviets to hold it from accelerating down the ramp. The cab of the truck took the brunt of the impact, but the generator was not damaged. As will be seen, the "standby" generator held that status briefly. The final sufferance was that one plane, albeit empty, returned from Yakutsk to the U.S. on three engines because the Soviets still maintained that one harness was defective.

Status and name change.

During the fourth flight episode, the CO tried on several occasions to visit the Soviet weather office in Khabarovsk without success. Their weather facility in Khabarovsk was actually never seen nor located by U.S. personnel. The only connection was a direct telephone line. It was learned that Soviet weather information was not encoded, the familiar international symbols being used. Major Vishvarka insisted that they receive copies of the codes used by the U.S. in transmitting weather data to Guam. Furthermore, the Soviet government requested plain language copies of all encoded messages transmitted concerning the work of the unit whether or not it contained weather data. The demands became so insistent that the CO asked that they be put in writing. The authorities responded two days later and put the demand in writing that they wanted to be fully informed of all business transactions. Again the CO reminded them that our only business in Siberia was to obtain the weather. Nevertheless, the CO turned the matter over to Washington and told the Liaison Officer that any further discussion would have to take place via official channels through Moscow to Washington. In the meantime, the CO would conduct his administrative business in cipher. The fact that the Soviets protested the use of the codes demonstrated to Capt. Cumberledge that they were copying all radio traffic emanating from the Weather Central. Even though the MOKO Unit had become unclassified on 18 September, the coding issue was not resolved.

To facilitate future dealings with the Soviets, Captain Cumberledge asked for a change in name from the "Advanced Base Unit MOKO" to "U.S. Naval Mission." In this way the CO reasoned that the unit could be associated with the advantages of the U.S. Naval Mission in Moscow. On 22 September a dispatch was received that the name of the unit had been changed to "Fleet Weather Central Khabarovsk" and the mission was that of a fleet weather central. The operational control of the Unit was under the Commander in Chief of the Pacific (CINCPAC) but the administrative control remained under Chief of Naval Operations (CNO).

By 26 September the radio teletype (Fig. 22) was placed in operation at the Fleet Weather Central Khabarovsk and Guam said it was ready to receive on its radio teletype. Ens. C. Birkett was not trained in the assembly of that type of equipment, but had the inherent technical skills to make things work. With the manual in hand, he and his radio technicians assembled the heavy racks of gear required for a 5,000-watt transmitter. (The assigned expert on radio tele-

Fig. 22 below. Specialist 3c(X) George M. Golovin at the radioteletype.

types who was to relieve Birkett made it to Fairbanks three months later, but was never transported to Khabarovsk.) The first test lit up the shack like a Christmas tree as a result of stray emissions from the powerful transmitter. The problem was solved by laying a network of wires 40 × 40 feet in the dirt floor to ground the equipment. A tape was punched in the radio shack about 100 yards away and transmitted to Guam. They reported a top signal strength of 5 and a low noise level. The CO was so pleased that he shared a bottle of his private scotch with Birkett and his men to celebrate their special efforts and success. The ruckus over codes still had not been resolved. It required another two weeks of negotiations to get a broadcast on the air.

Demand for ciphers.

The demand of the Soviets to turn over all ciphers and codes irritated Admiral Ernest King, CNO, rather strongly.[35] He di-

---

[35] Admiral King is said to have objected particularly to an alleged forty-eight hour ultimatum, according to both Brown (1962, p.83) and March (1988, p. 336). The official chronological history (p.3) regarding the demand for ciphers presents a different view: "The C. O. requested that the demand for ciphers be put in writing. This was agreed to. An answer to this demand could be expected in two days after the demands were put in writing." It appears that Capt. Cumberledge was merely stating that the Soviets would get a response within 48 hours after receipt of the demands in writing. No ultimatum on the part of the Soviets was implied.

rected the Naval Attaché in Moscow on 27 September 1945 to respond to the demand with the explanation that the ciphers were for administrative privacy. On 11 October 1945 General Alexei Antonov, Soviet Chief of Staff, responded with a harsh note to King that if the mission of the Fleet Weather Central Khabarovsk (Advanced Base Unit MOKO) was to transmit weather bulletins they should be done in the clear or in internationally accepted codes. He added that there had been no request to use ciphers and no permission had been granted. Furthermore, Antonov was concerned about the presence of two U.S. Army Officers, and especially Navy Lt. Worchel, in the MOKO Unit who had displayed special interest in "questions" that had no relationship to meteorology! The accusations regarding the "special interests" of Lt. Worchel are described below. Neither Major Kodis nor Major Bristor, the U.S. Army liaisons, were ever granted permits for travel outside of Khabarovsk.

Although advised by U.S. Ambassador W. Averell Harriman and others that if the weather data were that important, the U.S. Navy should send it in the clear. Admiral King responded equally harshly to General Antonov on 12 October 1945:

> I did not and still do not feel it necessary to request specific permission to use my own codes for administrative privacy in Khabarovsk. Soviet codes have been and are still used in U. S. territory without any question on our part and I feel the same practice is necessary for U.S. personnel in Soviet territory.

That exchange seemed to end the issue, but the fallout was substantial as events evolved. For example, Capt. Tikhonovetski dropped out of sight and Major Malakov, an engineer not trained in meteorology, acted in his place. The excuse was that Capt. Tikhonovetski had taken ill while on "business" in Moscow. Inquiries about his health received no response. It was not until the units were ordered to leave the USSR that Capt. Tikhonovetski reappeared on 16 December 1945.

4. CITY OF KHABAROVSK

City plan.

THE CITY SAT ON THREE MODERATE RIDGES that ran perpen-
dicular to the Amur River and terminated in steep cliffs. The
three main streets, about six lanes wide, were on top of the
ridges and were laid with granodiorite blocks and asphalt.
The adjoining roads were only four lanes wide and made of
dirt. The wide streets no doubt served as fire breaks because
of the large number of wooden buildings. There were few
brick buildings and the main government buildings were
faced with cement (Fig. 23). (A current Russian colleague
identified the architectural style as the "Stalin period"). Only
one building of quarried stone was seen. The tallest building
had six floors. Four apartment buildings were of modernistic
design. Again, it was difficult to distinguish "factories" from
dwellings. The industrial sections appeared to be in the extreme south and northeast sections of the city. Those areas
were restricted to all except workers. Nevertheless, it was easy
to identify a tank assembly plant, aircraft engine assembly
plant, and a farm machinery plant that used war-salvaged
materials. A cement and brick plant was visible; both are apparently rare in Siberia because of the lack of suitable raw
materials.

Amur River activities.

The principal street, named Karl Marx, led to the Amur
River where a seaplane port was visible with two U.S.-made
seaplanes (PBYs). These planes were transferred to Petropavlovsk when the river became frozen according to an officer
in the Soviet Merchant Marine. A flotilla was stationed at
Khabarovsk that consisted of a monitor with a single turret
(6"gun), several river boats about 150 feet long mounting 3"
guns, and three small, flat-bottomed, side-wheeler gun boats
with unprotected mounts for small caliber guns. Barges carrying coal, oil, boxed freight, and some passengers were
towed by tug. The side-wheelers struggled valiantly, emitting

Fig. 23. Karl Marx Street in Khabarovsk. From a 1945 post card provided by a Soviet citizen.

much smoke and noise, against the strong and turbulent current. River transportation ceased about the middle of December and did not resume until late April or early May. Commercial fishing was carried on for a salmon-like fish that was consumed locally. No fish canneries were identified, yet Khabarovsk was listed in descriptions of the city as a major packer of fish for a wide region. A railroad bridge crossed the Amur at the north end of the city. It was said that a tunnel also ran under the river that handled both rail and motor transport, but that statement was not verified. No one discussed how the Soviets were able to invade Manchuria with such speed and with such great numbers of troops with the limited number of river crossings. One of the Russian-speaking officers from the unit, Lt. Worchel, showing too much interest in the caves on the border of Manchuria, was arrested and declared *persona non-grata*. He was ordered to leave the country, but the order was rejected by CNO, so he was confined to quarters for the remainder of the mission. No doubt the Soviets were aware of his previous assignments as an Intelligence Officer and his two-year tour as Assistant Naval Attaché in the Archangel-Murmansk area as Harbor Master dealing with convoys bringing military supplies to the USSR.

Trans-Siberian Railway.

The railroad station, as in all cities, is an interesting place to people-watch. The Trans-Siberian Railway passes through Khabarovsk (see map, Fig. 1). All arrangements for passengers, baggage checking, and freight are made by the Chief of the railway station. Large numbers of people were continuously awaiting transportation; the cars were so crowded that passengers rode on the roof, on the steps, on couplings, and any other place that afforded a ride. Equipment was greatly in need not only for passengers but also for freight that was piled on sidings. As will be seen later the railway yard was a dangerous place in more ways than one.

Transportation within the city was by a few old buses with a capacity of about twenty people. Usually double that number managed to find a way to hang on. There was no transportation outside the city; one tried to hitch a ride on a military vehicle. Major Bristor recalls that people seeking transportation built fires close to the road to keep warm while waiting for a lift. One old women was picked up by the unit's jeep, and she was delighted for the ride and the high-ranking company. As a joke, Lt(jg) Uskievich asked her for payment as they dropped her off in the city. She obligingly reached into her sack and produced a large disk of frozen milk with the cream layer intact on the top. They refused the offer, of course, and she joined in the laughter after the American sense of humor was explained.

The Soviet trucks available were of light construction and often equipped with charcoal burners for power. Passenger cars were rare and a range of makes from the Soviet Zees to Japanese, German, and American. The heavy duty trucks were U.S. Studebakers. Equally rare were individuals on horseback or in a horse-drawn wagon.

Environs.

Due east of the city was a civilian airport (No. 7 in Fig. 24); the military airport was south- southeast of the city (No. 5 in Fig. 24). There was an emergency landing field near the village of Nekrasovka in addition to a Red Army garrison and a Japanese prisoner-of-war camp. The principal Red Army

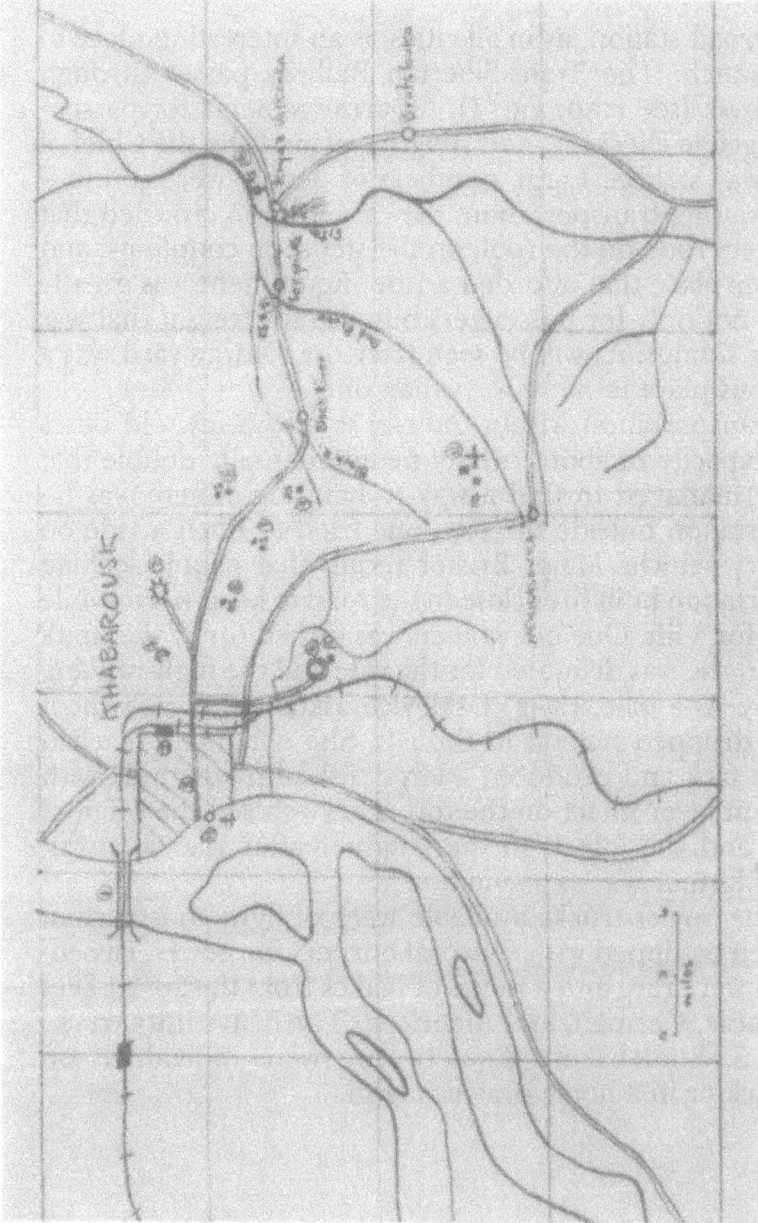

Fig. 24. Sketch map of the city of Khabarovsk and environs. The railroad bridge across the Amur River (1), Railroad Station (2), Karl Marx Street (3), Seaplane port (4) are identified. Those points noted with RS are various types of receiving or transmitting radio antenna. The crossed-picks symbol denotes a rock quarry at Knaz Volkonka (10). The cross marks a Prisoner-of-war (POW) camp (11).

Base (No. 10 in Fig. 24) was at Knaz Volkonka where there was also a major communication station. (The sketch map was enlarged in part from an aeronautical chart and the points of military interest plotted from memory immediately after leaving the USSR in 1946.)

Photographs of the region centered at Khabarovsk taken from satellites as early as 1959 were declassified by Executive Order on 22 February 1995. One DISP (Declassified Intelligence Satellite Photography) photograph taken on 30 May 1962 was of small scale and the pertinent areas partially obscured by clouds and snow. A Landsat MultiSpectral Scanner (MSS) image taken on 26 January 1973 was especially clear and is given in Fig. 25. The braided Amur River is to the west and north. The north-flowing Sita River passes through Knaz Volkonka, the isolated patch due east of Khabarovsk. The airstrip serving Khabarovsk and its nearby emergency field to the south are visible. The writer was unaware of the airfield readily discernible south-southwest of the town of Knaz Volkonka.

The people.

Most of the people were European Russians, Ukrainians, and a few Mongolians. About three-quarters had been evacuated from the west by order of the government. It was alleged that the Ukrainians were not trusted by the Soviets and were essentially replaced by Red Army troops in advance of their entry into the war with Germany. Even though the Autonomous Jewish Republic[36] was only 100 miles to the west of Khabarovsk, we met only one Jewish woman, a librarian. Only two Negroid people were observed in town. All who were willing to express an opinion greatly desired to return home but doubted whether that would ever be possible, even though promises had been made. It was indeed strange to see Mediterranean-skinned peoples in the severe climate of

---

[36] The Jewish Autonomous Oblast (Region) west of the Khabarovsk Krai (Territory) was set up in 1934 at the instigation of M. I. Kalinin, then Soviet President. It was essentially the creation of a home for the Jews of the Soviet Union. At first the Jewish migration to the region was voluntary, but a large number of Ukrainian Jews were ordered to move to the collective farms around Birobidzhan.

Fig. 25. Landsat image taken on 26 January 1973 of the city of Khabarovsk on the Amur River. The isolated village north of the eastern most airport runways is Knaz Volkonka, site of the MOKO base. The scale bar is approximately two miles.

Siberia. According to the local town's people, the major new city north of Khabarovsk on the Amur River, Komsomolsk, was populated by Ukrainians who worked primarily in the aircraft factories there. (Komsomolsk was named for the members of the Young Communist League, the Komsomol, who were the principal builders of the new city in 1932.)

Social life.

Khabarovsk had a theater, three movie houses, a dance hall, and a circus ground. The schools included a pedagogical institute, nurses school, medical institute, railroad engineering school, and a law school. (It was pointed out that the nearest college for science was in Novosibirsk.) Only one church was known to be in the city. Even though attendance was frowned on and religious holidays banned, older people were willing to take the risk. Apparently Section 124 of the Soviet Constitution guaranteeing freedom of conscience and of worship for all citizens was interpreted in a unique way by the authorities. To avoid persecution, the family icons were usually hidden from the authorities. The paintings were in the classical style of the eighteenth and nineteenth centuries and were indeed artistic works of great beauty. One officer from the unit, Lt. Worchel, persuaded several of the local families to part with their icons (=ikons) on which are depicted sacred personages or events of deep religious significance. Unfortunately, the icons deteriorated during Lt. Worchel's many moves in and out of high humidity environments after the war and had to be sold.

Parades, requiring the assembly of a large number of people, were attended only with a special pass and home gatherings were exceedingly rare. On government approved holidays parties were held that were focused on wine, food and conversation. Class distinctions were quite evident as measured by the privileges and passes granted. People with positions of responsibility dressed, lived, and ate better. Most women, however, wore horsehair boots (valenkis, Fig. 26), heavy long underwear, knee-length skirts, a quilted jacket, and a scarf or knitted cap on their head.

A highly prized perk was a chauffeured automobile. But assignment to a large dacha, a private country house, was

Fig. 26. Felted boots (valenkis) made of horsehair. For a child, 9 inches from heel to toe and 13 inches high.

usually shared with one or two other families. The privileges seemed to be directly related to membership in the Party, a highly restricted and select group. Needless to say, the distinction between officers and enlisted men was very rigid indeed. One had to conclude that Soviet society, presumed to be classless in 1945, had more layers than the rigorous distinctions in an East Indian Society! In short, the divisions between the Muzik, the peasant, and the Commissariat, a Party member, were multi-fold.

Utilities.

Water was obtained from the Amur River and filtered, according to a Red Army doctor, who recommended that it be boiled. An enlisted man said it was not boiled before the war, but they feared it had been polluted by the Japanese during the war. Most people drank tea to make the boiled water more palatable. The water was pumped to several central locations where the population could draw their water. The outlying areas were served by a community well. Running water in a home was rare and few toilets were operated with running water. Bathing facilities were limited and small metal tubs were seen in the larger homes.

A central steam-powered, electrical generating plant furnished the city with 220-volt, alternating current. The peak load was between 5 and 9 P.M. based on radio volume and light intensity. Apparently a second generator was cut in at 5 P.M. to handle the load. No high-voltage transmission lines were observed, so it is not known how the local communities were served. The local village of Knaz Volkonka had lights for a couple of hours in the evening. The frequency was around 50 Hz and could not be used by the unit's equipment designed for 60 Hz. It was suddenly noticed that the lights were burning in the village a little longer than customary, and Birkett discovered that someone had tapped into the unit's generator! Needless to say, the unauthorized line was removed, but the perpetrators were given an "A" for effort.

Wood was the principal fuel for heating. A local wood camp was operated by the Red Army about 15 miles southeast of Khabarovsk. A civilian directed the cutting of trees ½ to 1½ feet in diameter. The trees were typical of the subpolar taiga; pine, birch, and oak were seen. The long cold winters and low precipitation had no doubt limited their growth. As the MOKO Unit discovered first hand, the cutting of wood was highly controlled in regard to where and how much was to be cut. A small amount of very soft bituminous coal bordering on lignite was used to heat a few homes and fired the larger heating systems. The Soviet liaison officer said the coal came from the town of Blagoveshchensk, some 500 miles up the Amur River. Most buildings were heated with individual stoves or a hot water system on each floor. The Russian stove is a very efficient device and will be described below in detail (see Fig. 27). Government buildings were usually kept at low temperatures; the occupants wore their overcoats indoors.

Sewage was simply disposed of in the two streams running between the three ridges on which the city is built. Septic tanks were rare; outdoor privies were common. It seemed as though the outhouses once built were never cleaned. As every farmer in the northern parts of the U.S. knows, use of the outhouse at temperatures near −40°F is kept to a minimum and carried out with great speed.

Radio receivers were not seen in private homes, but most had a loud speaker that furnished programs from a central station. It was mandatory that the loud speaker be kept on at all times in order to receive government bulletins and directives during the war. (Members of the MOKO Unit were able to pick up broadcast stations from Moscow, Irkutsk, Vladivostok as well as Khabarovsk.) Captured German radios were on sale in a general store, but a permit was required to buy one. It was not considered a good policy for an individual to possess a radio receiver. One officer learned that before

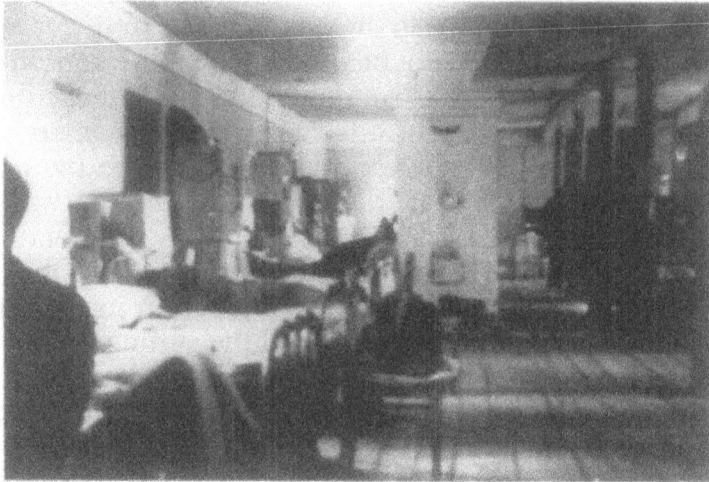

Fig. 27. Interior of quarters for enlisted men (Fig. 8) illustrating one type of Russian stove (pechka).

the war some cities were permitted to distribute receivers. These were collected, however, for the duration of the war. Major Bristor tells of a man now living in the U.S. who had been arrested at the age of 14 in Bucharest for possessing a short-wave radio. The boy's father was also arrested, but did not survive the gulag. The boy managed to stay alive for ten years at work in the mines of Magadan before being released.

Telephone communication was not very dependable and the connections noisy because of other conversations, generator noises, music, and radio signals. All telephones used by foreigners were monitored to the point that the conversation would be interrupted if the speaker was not clear or the identity of the talker required verification. Long distance calls to Vladivostok required 10–30 minutes to complete the connection. Calls to Moscow were by appointment only and then only between 4 and 6 A.M.[37]

Telegraph service was even less dependable. Official

---

[37] There is a 7-hour time difference between Khabarovsk and Moscow, that is, 4 A.M. in Khabarovsk is 9 P.M. the previous day in Moscow—not a useful time to reach a working office! No long distance calls were ever completed to the MOKO Unit unless arrangements were made in advance to receive the call in the Soviet liaison officer's office.

dispatches sent to the U.S. Naval Mission in Moscow or to the U.S.A. were never received. Personal messages in Russian were received, but delayed four to seven days. Messages in English text required twice that time. About 80 percent of the messages to Vladivostok got through, but one could not be in a hurry to pick them up. Transferring a telegram from Khabarovsk to Knaz Volkonka took 2–4 days. In short, a third party had to review every message. Experience with cablegrams to the U.S. fared about as well; about 75 percent were delivered. Prepared replies required up to two weeks.

5. LIBERTY IN KHABAROVSK

Friendly Reception.

EVERYONE WORKED strenuously to set up the MOKO Base, but there were times when specific personnel would be relieved from their duties because of lack of materials, delays in obtaining permissions from the Soviets, and the need for relief from the frustrations of working under difficult conditions. When available, the jeep and truck were used to take personnel for brief visits into Khabarovsk some 23 miles away (See Map, Fig. 24). (Because of the poor road conditions and extreme cold it was necessary to send two vehicles into town as a safety measure.)

A large number of men went in on a Sunday afternoon early in September when the town was crowded and the weather was reasonable. The initial reception was exceptionally friendly and some even received a handful of flowers in a gesture of friendship. There was considerable interest in simply talking to and touching Americans even though the conversation was limited by a mutual lack of language skills. In the excitement of exchanging small gifts, one young lady mistook a brightly wrapped bar of Ivory soap for a candy bar much to the amusement of the surrounding crowd. Strolling up and down the main avenue and drinking "fruit water" from street stands was the principal activity. Artisans were available to pencil sketch, paint, or cut out a silhouette of the subject (Fig. 28) for a small fee. There were at least five parks,

Fig. 28. Silhouette of Lt. Yoder cut out in a few minutes by an artisan on the main street of Khabarovsk.

each of which had a "walking circle" where friends could walk hand-in-hand, round and round, while talking. It apparently was *the* place to be seen socially.

Internal security arrives.

The second group to visit Khabarovsk had a different experience. Members of that group were obviously being followed by the NKVD[38] and none of the townspeople took the risk of talking to foreigners. The dress of the plain-clothed NKVD was unique and the uniformed men had green hat bands that were clearly identifiable. On occasions when one was lost in the city, the easiest way to obtain help was to just turn about and walk back to the agent for directions. One of the tricks soon devised to elude the NKVD was to start out in groups of three. Because the security men always worked in pairs, they could only follow two of the three if they suddenly dispersed in a crowd. In this way one person always got to walk freely without surveillance. There was no serious purpose in the trick except to beat the system.

Movements Observed.

About the middle of October, the Executive Officer was questioned about the activities of the CO in Khabarovsk on the preceding day. Apparently the tail on the CO slipped up, and there was a gap in the record. The Executive Officer explained that their mission was to obtain weather for the fleet, and he had no secrets and nothing to hide. Although told by the Soviets that it was not necessary, the CO felt it was his duty to make official calls on the mayor of the city and the Chinese Consul recently established there. Major Vishvarka replied that it was his job to keep the American Captain informed of the "Russian manners of doing things." He understood there was some difficulty in getting into the Chinese Consulate because a member of the NKVD posted there de-

---

[38] The letters NKVD translate into the People's Commissariat for Internal Affairs, that is, the Soviet police agency responsible for the labor camps and for internal security in general. Their broad investigative powers were primarily focused on political offenders.

manded a pass to get in. The Captain said this practice was
unnecessary because the mayor himself had said ". . . this is
the Soviet Union and you are free to go where you like and
the party had no right to stop and request an admittance pass
to the Chinese Consulate."

   Not to be dissuaded, the major bluntly asked the captain if his business was accomplished successfully. He explained that he had not gone there for business reasons but
merely to pay his respects to the Mayor and the Chinese Consul. Equally blunt, he asked the major if he wanted to know
whenever the CO went to town and where he was going. The
frontal assault worked, and the major replied

> that he was not interested so much in where the captain went
> while in town for whatever business he had, but only in so far
> as where the car went when it was in town. In this way if he
> knew that the car was going to be over by the hospital, or over
> by the park, or over by the telegraph office, that if any me-
> chanical trouble with the car should occur that the major
> would be able to get the assistance of the captain much faster.

At the end of the sparring match of words, the major stated
that the Captain could do as he pleased and did not expect
to be informed, but many of the difficulties encountered
could be forestalled or avoided if he were informed (CO Report No. 24, 18 October 1945). It seemed clear that the Major
as Liaison Officer was responsible for knowing every movement of the CO of the Weather Central.

Saturday night dance.

When the weather was appropriate, the Red Army band
played at an outdoor dance pavilion. The raised wooden
floor emitted clouds of dust as the dancers performed polka-
like steps. Lt. Starke recalls that the dance floor was patched
with boards nailed over the rotted sections, creating another
hazard for the fast moving dancers. For reasons unknown,
the girls danced with girls and men with men, in the majority.
The author was invited to dance by a Soviet colonel[39] and

---

[39] Apparently the senior Soviet officer present was required, as a matter of courtesy, to invite the senior American officer present to dance.

every effort was made to keep up the pace. The polka is rather energetic, and in parka and boots it becomes a significant workout!

Some of the more exuberant Soviet enlisted men performed the Ukrainian dance called Hopak. In a squat position with arms folded, the legs are alternately extended to the rhythm. It is indeed an exciting performance with many variations including high kicks, clapping, and whirling. One has to be quite limber before trying such an energetic dance.

Theater.

The author was fortunate to have seen two plays, a ballet, and the circus. Not knowing the language restricted one's appreciation of the plot, but the presentations were very professional. Both plays were musical comedies called "Columbine," and "Perikola." The ballet was a vigorous presentation of "The Red Sailor's Dance." The circus was enjoyed by all, requiring no language facility. The acrobatics were spectacular and the dancing black bear was the highlight of the presentation, according to audience response.

Two negative aspects of the theater remain vividly in the memory. There was no heat in the building, so heavy overcoats had to be worn. As the closely packed audience gave off body heat in the closed room, a very disagreeable odor developed. Dry cleaning was apparently not available under wartime conditions in that part of the world. This smell was compounded by the characteristic aroma of smoke from Soviet cigarettes, presumably of Turkish origin.[40] The intermissions were a welcome relief to get fresh air, particularly after use of the so-called toilet. Imagine a dimly lit room with a wooden trough on one wall. The other walls were used as support for eliminations that accumulated against the wall. The trick was to get from one side to the other without stepping in it. Because of the low temperature, the aroma was

---

[40] Soviet cigarettes were sold in rigid, flat, flip-top boxes. The cigarette, 8 cm long, is only three-eighths tobacco and the remainder a hollow paper tube. It certainly reduced waste of tobacco compared to American cigarettes, which were highly prized. On the open market, a pack of American cigarettes was traded, for example, for a new pair of children's horsehair boots (valenkis, Fig. 22) as a souvenir. Trading American cigarettes was not appreciated by smoking colleagues, but it was no sacrifice for a nonsmoker!

minimal. Such bodily functions were not important to the culture and were essentially ignored in the local society. On the other hand, there were several American traits (e.g., whistling, boisterous laughter, chewing gum, aggressive and rude behavior) that were considered "Ne Kulturni" (not cultured) by the Soviets.

Museum Visit.

Only one visit was made to the local museum. I cannot recall whether it was attached to an institute or was maintained by the city. It was a useful place to go to learn some of the history of the region, to associate objects and Russian words, and to see products of the local craftsmanship. Of special interest was the display of rocks and minerals of the area. Samples of the rocks outcropping nearby were displayed. Ores of tin, molybdenum, and antimony revealed that mining was carried out nearby. Local fossils of substantial size were exhibited. No one else came into the museum during the visit, and the curator stayed at her post, so it was possible to wander freely among the exhibits.

Books valued.

It was entertaining to try to deduce what took place in the various buildings along the streets. The principal street had two libraries, and four book shops. One book shop specialized in military books, another on teaching aides, and another served as a clearing house for textbooks. Reading was considered important to the Russians and books were one of the cheapest items to buy. In order to learn the language I bought several teaching aids for children such as an alphabet strip, fairy tales (e.g., Tolstoy's "Three Bears"), a geography atlas for middle school, and a pocket dictionary. The geography book turned out to be more informative than expected. The states of Latvia, Estonia, and Lithuania[41] had already been incorporated into a map of the USSR dated 1944. Fur-

---

[41] The annexation of Lithuania, Latvia, and Estonia by the USSR took place in 1940. That action was never recognized by the U.S. government.

Fig. 29. "Schoolmaster, who did not   Fig. 30. "Benzene test"
wish to praise the Germans"

thermore, locations were given on the resources charts for
uranium mines then unknown to the U.S.!

One pamphlet purchased deserves special mention because it exemplifies the attitude of the Soviets toward the
Germans. The large (1×1½ foot) presentation of 15 litho-
graphs simply titled "They" [i.e.,The Enemy] by S. M.
Chexov (1943)[42] depicts atrocities attributed to the Germans
(Figs. 29–34). Because of their size it appeared as though
they were to be used as posters.

The only small billboard posters observed in the city
illustrated illegal voting by fat capitalists. A Soviet officer explained that the poster was concerned only with events in
England! It was a weak effort to avoid application to the U.
S., but it at least showed that they were conscious of our sensi-
tivity to remarks downgrading democratic principles. The

---

[42] The original pamphlet has been deposited in the Hoover Institution on War,
Revolution and Peace, Stanford University, Stanford, CA. A Finnish friend who
fought first against the Russians (30 November 1939–13 March 1940) and then
with the Russians against the Germans (after 3 March 1945) attested to events
similar to those portrayed in Figs. 31 and 32.

Fig. 31. "Still, what remains"

Fig. 32. "Ice pillars"

Fig. 33 "She 'concealed' her warm be-
longings"

Fig. 34. "Ours"

large billboards were reserved for Communist slogans regarding work and victory.

Absent were the customary news stands; however, pages of *Pravda* (Truth), the official newspaper of the Communist Party established in 1912, were displayed in billboard fashion. One officer obtained a copy of *Pravda* about mid-September when the Soviet citizens were informed about the "atomnaya bomba" used on Japan. The local four-page newspaper distributed in the Khabarovsk region was called the "Pacific Star."

Cheese caper.

One store specialized in dairy products. Well past the lunch hour, the prospects of a cheese sandwich was inviting. The line was not excessively long and the women in their babushkas (head scarves) carrying fish-net shopping bags were congenial. After reaching the counter, fifty rubles (10 US dollars) were laid on the counter and attention was directed to a large wheel of what appeared to be cheddar cheese. A snicker arose from the group behind as the clerk brandishing a cleaver-like cheese knife sliced off a sliver with a flourish. The thinness of the slice rivaled the perfection of a microtome. The snickers grew into laughter as the slice was placed in the hand. Stuffing the slice into the mouth, about facing, and marching out the door, the author learned another lesson in price variability of products around the world. Prices were not marked on any of the items because of the drastic price changes that were highly dependent simply on availability. The economic system seemed familiar for that one moment.

Similar difficulties were encountered in buying candy. Lt. Starke laid a few rubles on the counter and pointed to the "penny" candy pieces in the tray. The salesgirl carefully weighed out five pieces, each individually wrapped, on a beam balance. Because it was a little light, she cut another piece in half to bring the weight up to the value paid. Apparently there was a shortage of coins for making change. Most of the kopek (100 kopeks = 1 ruble) coins obtained at other shops were dated in the period 1935–1943. The chocolate candy, labeled Siberian Bear, was made in Moscow by the firm Pom-Fronm that is believed to be no longer in business.

The hard candies had also been made in Moscow by the company P. A. Babaeva, which my Russian colleagues say still exists today.

"Casual encounter."

On a liberty early after arrival, another officer and the author were approached by two young ladies (Fig. 35). Both were attending the local institute and had been studying English. Lyda Mircotan spoke English very well because her mother taught the subject. Fortunately, the other officer (Ens. Bowden ?) could speak Russian and got along well with Luba who was not as proficient in English as Lyda. After long strolls in the park it was agreed to meet the next time we were on liberty. A note was to be placed in a specific tree as to the time and place to meet. It was noticed that the NKVD did not follow us, so it did not take long to deduce what the purpose of the encounters were. The discussions were restricted to life in the U.S. and our families. In that environment, the word picture of the U.S. does come across as a fairyland. The idea of refrigerators, cars, air travel, individual houses, the wide variety of foods, telephones, and radios as common place was not believed at first. Apparently every word was relayed first to her mother and dismissed as pure propaganda. Lyda nevertheless absorbed a rather detailed picture of U.S. family life and began to express interest in coming to the U.S. Her knowledge of our style of democracy was exceptional, and the author was not always prepared to deal with all the apparent negative aspects for which she had been well rehearsed. The finer points of voter fraud, disenfranchisement, pork-barrel bills, income-tax loopholes, inspection and permit corruption, contract kickbacks, payoffs, bribes, and other practices were spelled out in great detail by Lyda. The mother noticed a change in her daughter as a result of these discussions, and invited me to their apartment for dinner.

Because of the father's position, the apartment consisted of two rooms on the top third floor. One room was for sleeping and the other for sitting and eating. Clothes were stored in chests and cabinets. The perpetual loud speaker was in the sitting room. Cooking was done in a small commu-

Fig. 35. Lyda and Luba.

nity kitchen at the end of the hall. The toilet was also communal. It was the custom to bring a gift of equal or greater value
than the dinner. For this purpose a suitable collection of
American canned foods was provided. The traditional round
loaf of bread and dish of salt was on the table, a sign of
welcome.

It was subsequently learned directly from her mother
that Lyda was being prepared to enter the Party and her eval-
uation of the capitalist system was the last test. The mother
pleaded with me not to fill her head with such "nonsense"
about American democracy because it was ruining her
chances for membership. No amount of persuasion regarding the truthful nature of the description of life in the U.S.
was effective. The father, who was head of the local cooperative farms (Kolkhozes), did not understand English and did
not enter into the discussion. Her only brother was interested
in agriculture, but had been assigned duties as a radioman
because that was what was needed. It was most unlikely that
this particular family discussion was ever reported to the authorities. Such discussions were usually held while walking in
the park and not in the house. Because each age-group of
children was organized in a political sense and met separately
from the other age-groups, it was possible to cross check the

stories innocently divulged by the youngest members of the
family. For this reason, for example, complaints about the system never got discussed at the family dinner table.

Enhanced by the fact that the mother taught at the local
Institute, there was great pride in Russian culture within the
family. On one occasion a thick volume of Russian Poetry was
presented as a gift and portions read aloud. The flow of the
language was most pleasing even though the words were not
understood. A parting gift consisted of a beautifully bound
copy of the collected works with an extended biography of
the romantic poet M. Y. Lermontov (1814–1841)[43] edited by
Chuiko (1894) printed in old Russian.[44]

Lyda wrote one letter after the unit returned to the U.S.
It was apparently hand carried by a friend and mailed in
another envelope[45] from a small town in Pennsylvania. She
had passed her examinations at the institute, but no mention
was made of her becoming a Party member. As a reward for
her scholarly success she was allowed to visit her brother in
Vladivostok where she became friends with the U.S. consulate through her previous association with Irene Matusis, assistant to the Asst. Naval Attaché, when she visited
Khabarovsk. Enclosed in her letter was the post card reproduced in Fig. 23. Because of the rise of anti-communism in

---

[43] Lermontov is considered the foremost Russian romantic poet. He wrote poems
at the age of 14 in the vein of Lord Byron, but soon developed an emotional style
in describing the political events of the day not always acceptable to the aristoc-
racy. After a brief period of exile in the Caucasus, he was allowed to return to St.
Petersburg and was assigned to a local regiment. He was brought to trial in 1840
for a duel with the son of the French Ambassador, and again sentenced to exile
in the Caucasus. Lermontov distinguished himself in heavy fighting in the battle
of Valerik River, only to die in a duel with a fellow officer. His heroic treatment
of socially relevant themes promoted a new direction for Russian literature.
[44] New spellings of Russian words were decreed by the Council of People's Com-
missars in 1918, and were used in all official Russian writings thereafter except
in those of the Academy of Sciences who adopted them later in 1924. Some letters
in the Cyrillic alphabet in old Russian that were dropped are diagnostic of the
ancient East Slavic language and the golden age of Russian literature.
[45] The inner envelope had five, uncanceled, Russian stamps attached: four 10-
kopeck stamps illustrating a female factory worker (Scott # 616) and one 30-
kopeck stamp showing an aviator with a propeller airplane (Scott # 736). It was
surprising that common stamps, issued in 1938 and 1939 respectively, had been
used. Many commemorative stamps had been issued during the war praising
their victories and heros, but apparently they were not released in Siberia.

the US, no attempt was made to reply to the letter or communicate in anyway with the Soviets. The Federal Bureau of Investigation (FBI) was kept apprized of any contacts remotely related to the mission.

Other casual encounters did not evolve further—either out of fear, potential compromising situations, or because of language difficulties. For example, Lt. Starke attended a one-ring circus with a lady he met one evening, and they made plans to meet again on his next liberty. When he visited the address she had given him, there was a reluctance on the part of the occupants to talk to him, to have any knowledge about her, or claimed that she no longer lived there. On another occasion he dated a female Russian officer assigned to accompany the group to Vladivostok who invited him to visit a military installation. He declined that potential confrontation, but agreed to attend the theater with her that evening. She appeared in civilian clothes, which she confessed were "relief" clothing from the U.S. Nevertheless, they saw the play from a box directly above one end of the stage. Again, the test of not being followed indicated that there was a purpose behind their meeting.

Liberty restricted.

On 27 September liberty for the unit was to be limited to two days per week; however, it is not known what precipitated the restrictions. Protests to the Liaison Officer only resulted in the admonition that this was considered a law and "they were to obey the rules." The approaching colder weather was more effective than the rules in limiting visits to town where little comfort could be found. Apparently the unit was overtaxing the NKVD and there were insufficient men available to maintain the required surveillance. It was evident that too many groups were involved in surveillance as events of 17 October revealed. Six enlisted men while in town on leave from the unit were accosted by two Soviet officers. Apparently the officers were insisting that the men produce some identification papers, but none of them understood Russian. Suddenly a third Soviet officer ran up behind one of the sailors, threw his body into the sailor's back in an effort to knock him down. Using good sense, the sailors thought better of settling the

matter with fists and quickly walked off to the park where
their vehicle was parked to take them back to base.

After reporting the incident to the Liaison Officer, the
issue of greatest concern to him was whether the Soviet officers wore red arm bands on their sleeves. Apparently there
were two police organizations in town; one, the local city police and the other, the military police. The Liaison Officer
suspected that it was the military police that were involved
and promised to have the three officers dealt with severely.
He assured the CO that the safety of the Americans would be
maintained, and that such an incident would never happen
again. Fortunately, there had been no bodily harm, but it certainly made the case for providing proper identification
papers to all personnel. By 25 October all personnel had
identification papers issued and certified by the American
Embassy in Moscow. In turn, Major Vishvarko had contacted
all organizations in Khabarovsk and thoroughly instructed
them with regard to the presence of Americans in the area.

6. BASE ACTIVITIES

Horseback riding.

ONE OF THE RUSSIAN-SPEAKING OFFICERS became friends with a member of the local Soviet cavalry stationed on the Sita River. Arrangements were made for six of the unit's officers to go on a pleasure ride with about a dozen cavalrymen. The commanding officer of the troop could not come but volunteered two of his best horses. All the animals were spirited, well disciplined, and trained in military maneuvers. The horses were relatively small, a hand or so smaller than U.S. horses, and their hair was quite long.

All of the group were comfortably mounted and there was a discussion about the direction in which to go. As the discussion proceeded in Russian the word "Stoyachee" was mentioned regarding the use of the horses in crowd control. As a demonstration, the command "Stoyachee" was given and all the horses immediately stood on their hind legs and flayed their forelegs. Not knowing that the word meant "stand up," the author, sitting the saddle appropriately with heels down, promptly slid off the rump of the horse to the great laughter of the group. The horse's panic was equal to the rider's embarrassment! It was a difficult way to learn the language, but the lesson had a certain lasting value after the pain subsided.

The ride consisted of periods of trotting lasting about 30 minutes and of walking for 10 minutes. At that pace the animals could do up to sixty miles a day! The animals—and the riders—were greatly invigorated by the ride despite the cold. It was a pleasant way to see some of the rolling countryside, too.

Volleyball.

The summer camp had a volleyball court. The game provided just the right amount of exercise, even at below zero temperatures. It did not take long for a challenge match to be arranged with a nearby Army post. On the day of the match the temperature was $-18°$F and no one needed encouragement to keep moving. The Soviets played in their undershirts as the rest of us turned blue. Needless to say, they beat us badly in all three games played. Their teamwork was excellent and the ball placement most accurate. Fortunately for us there was no request for a return match.

Chess—a devious diversion.

As a result of the U.S. refusal to turn over all ciphers and codes, the Soviets tried another tactic to prevent the transmission of coded messages outside the USSR. (They continued to retain full control over any telegrams, telegraph, or teletype messages within the USSR). A jamming station was erected by the Soviets well within sight of the compound down line from the unit's V-antenna oriented by necessity toward Guam. Attempts were made on many occasions to have chess matches with the radiomen stationed at the jamming station. Although the unit had some experienced chess players, the Soviets were indeed skilled at the game. Efforts to divert the radiomen from their assignment of jamming the weather broadcasts were not always successful, but the participants greatly enjoyed the games. Eventually the V-antenna was abandoned and a simple, single wire with a central tap was used according to Birkett. The reasons why the Soviet attempts to jam the unit's transmissions were unsuccessful are not wholly clear at this distant time. Either their power was inadequate, the wrong frequencies were transmitted, or the polarization was misdirected. On the other hand, the one leg of the V-Antenna used may have been oriented directly at the jamming station, which was, therefore, at a null point of the transmitting lobe of the dipolar antenna. For whatever reason, the transmissions of the unit were received in Guam loud and clear.

Russian movies.

Two movies were provided for showing at the base. One was the four-hour Russian epic called "Ivan the Terrible," Part I, issued in 1944, and produced by Sergei Eisenstein with background music by Prokofiev. The history lesson was most interesting, but the palace intrigues were not well understood. As the first czar (Caesar) during the middle 1500s, Ivan's military campaigns against the Tartars added non-Slav territory to Russia. His wars against the Poles and Germans were waged in order to break through to the west and establish trade with England. He is credited for establishing the first national assembly in Russia.

The second movie "Alexander Nevsky"[46] contained battle scenes, which were especially realistic, depicting a Russian victory over the Germans in April, 1242. The Teutonic knights were driven on to the ice on Lake Peipus southwest of St. Petersburg and fell through under the weight of their armor. It was a unique scene in the epic. The battle was apparently re-enacted with considerable enthusiasm by the actors, and there was discussion by the Russians about the number of real casualties during the filming of the battle scene. Even though the movie is cited for its innovation, the attempts at artistic presentation through curious angle shots detracted from its enjoyment. All in all, there was too much to absorb with the many changes of reels, and most of the group departed before the end of the movie.

Fireplace constructed.

The Russian pechka is a very efficient heating and cooking stove. One type is displayed in Fig. 27. Within the box-like structure is an open maze of bricks around which the heat and smoke rise from the oven-like fire pit. In other larger types the stove and chimney are built in steps so that cooking can be done on the first level and successive raised platforms each consisting of an open maze of bricks, on which one may

---

[46] The film directed by Sergei Eisenstein with the musical score by Sergei Prokofiev was released in 1938, but was interpreted by the Germans as a political deterrent to a suspected invasion of the USSR by Nazi Germany. After the German-Soviet nonaggression pact of 1939 was signed, the film was withdrawn from circulation. Nevertheless, the defeat of the Teutonic knights by Nevsky made him a national hero, and the film redistributed after the Germans invaded the USSR on 22 June 1941.

It is ironic that between September 1941 and March 1943 Lake Ladoga, (northeast of St. Petersburg) when iced over was the survival link for the besieged city, then called Leningrad, to the rest of the country. Wounded and sick were transported out and supplies brought in over the ice.

sleep. A fire is built in the evening for cooking, the bricks within being heated, and the platforms became comfortably warmed for sleeping. The older people get the lower platform and the children the upper levels.

Unfortunately the efficient and wood-conservative pechka lacks all the romance and the intensity of heat of an open fire place. To remedy that deficiency an open fire place was designed and built in the officer's mess hall. A design was made solely from memory of a fire place in the author's home. A local bricklayer and a load of bricks were obtained through the liaison officer. A hole was cut in the wall that in itself was of unusual construction. Two walls were discovered and the space between was filled with sawdust for insulation. One could not complain about the cost of mortar. The bricklayer chopped up the local dirt outside the building, made a slurry with boiling water in the pit, and scooped out the "mud" with his bare hand. With the hand as a trowel the "mortar" was spread on the brick and put in place. It only took a few minutes for the "mortar" to freeze and the brick locked in place. The bricklayer was most efficient but his hand was stiff as a board at the end of the day.

It seemed appropriate to look for some local natural stone to face the fireplace. The author being trained as a geologist set out to look for suitable rocks. After bending over to pick up a few rocks to check for suitability, the NKVD appeared from nowhere and an arrest was made. It took the CO a day to persuade them that the rock collecting had a legitimate purpose and there was no other interest. That experience terminated any further attempts to collect rock specimens.

The damper was not of the best design, but a metal deflector made of tin cans served to route the smoke up the flue. You can be assured that the author has paid more attention to the construction of fireplaces since then. The open fireplace consumed large amounts of wood, but brought considerable enjoyment on the coldest evenings. Until the design was corrected, it was difficult to sing joyful songs with tears in ones eyes from the smoke. It certainly provided some relief from the daily frustrations of getting the weather central operational.

One roaring fire almost resulted in the burning of the building. The stone apron in front of the fireplace was not adequate and the sawdust insulation under the floor boards began to smolder. Fortunately, it was discovered in time to prevent a disaster. Major Vishvarka was disgusted with the entire project, especially with the excessive "waste" of wood, but it did not seem to deter him from enjoying the fire.

Hunting.

Captain Cumberledge and Lt(jg) Uskievich were invited to take part in a hunting trip organized by the Liaison Officer Major Vishvarka on Sunday, 11 November 1945. (This date was initially proposed for the first landing of troops on Japan—the choice of Sunday in remembrance of Pearl Harbor Day and Armistice Day of World War I no doubt had its implications.) The game was Siberian geese and the appropriate weapons were provided. The shells, containing excessively large pellets (buckshot for riot control?), were made of brass and were recovered for reloading. To retrieve the game shot down (in lieu of dogs), the Asst. Liaison Officer stripped to his waist and plunged into the Amur River and retrieved the birds. Uskievich thought the mosquitoes were of sufficient size to qualify as small game: They thrived on citronella, a common insect repellent. Enough birds were bagged for the forthcoming Thanksgiving celebration. In spite of their many confrontations, one got the impression that the Captain and the Major enjoyed some measure of personal friendship.

Ice Skating.

The frozen Sita river was a constant invitation for ice skating. The two Army officers bought a couple pairs of old fashioned ice skates in Khabarovsk that they could bolt on to the high-cut army boots. Such skating expeditions were usually short-lived because of the penetrating cold. In an effort to use up his non-convertible rubles, Lt. Starke also bought a pair of skates for only a few rubles, but they required permanent riveting to your shoes.

Celebrations.

On 28 October 1945 a Navy Day[47] party was held in the evening. The enlisted men celebrated in their Recreation Hall after a special dinner in the Mess Hall. The officers entertained in their "Club" with dinner followed by a movie and dancing. The fifteen official guests included three members from the Chinese Consulate, officers from the Khabarovsk Military District, the Meteorological Service, local Army Garrison, the Amur Flotilla as well as members of the Liaison Staff. Cmdr. Butow emphasized that having Soviet guests at the base would "help cement mutual friendship and understanding." According to Major Vishvarka, Capt. Cumberledge and Cmdr. Butow made the guests feel at ease. Especially helpful in making the party a success was Klopovic[48] and a personable young lady, Miss Irene Matusis[49] who served as translator for Cmdr. Roullard in Vladivostok. For unknown reasons, Cmdr. Roullard was not given a travel permit to attend.

The party was organized by a committee who prepared the list of foods and beverages as well as developed the entertainment program. The Liaison Officer obtained the suggested supplies at the expense of the unit. The request to invite ladies from the city was not responded to until three days before the party. The result was that the Major had to

---

[47] The US Navy was first established on 13 October 1775 by the Second Continental Congress. It essentially dissolved with the sale of the last ship in 1785. A permanent Navy Department was established on 30 April 1798. Today the Navy's birthday is celebrated on 13 October. The reason for its celebration on 28 October is not known, however, the preferred date remains a controversial issue.

[48] Klopovic was very fluent in Russian and was able to generate humorous phrases at the party. In addition, he kept the waitresses in the mess hall in stitches. For example, he used the Russian word for "puff pastry" that was similar to the word for "young girl" in calling a waitress.

[49] Irene Matusis was born in Moscow of Russian and American parents. After her parents moved to the U.S.A., she served under Cmdr. George Roullard, Asst. Naval Attaché then in Moscow. She accompanied him to Archangel and Vladivostok, but when his tour of duty was up, she was promptly arrested by the NKVD. It was concluded from the fact that she was allowed to associate with foreigners that she had been recruited by the NKVD. For this reason no sensitive matters were ever discussed in her presence. Nevertheless, she was also of great assistance to Capt. Cumberledge. To this date her fate remains unknown to her American friends. Capt. Cumberledge states that Irene Matusis was an American citizen with a US passport, but was unable to get a Russian exit visa.

agree to extended periods of leave and provide transportation so the men could go to town to extend invitations. Transportation was arranged for the guests.

Thanksgiving was celebrated by the unit with a group party and a dinner of goose was prepared in the holiday style. Guests included the U.S. Consul General in Vladivostok and his wife, who had been visiting the Weather Central, the Chinese Consul General and the Assistant Consul in Khabarovsk, and the Liaison Officer and his aide. Entertainment consisted of clever skits, songs, dialogs, and a game of charades that were produced jointly by the officers and men. Ens. Bowden and Storekeeper (Disbursing) First Class Hazzard spent Thanksgiving in Vladivostok as guests of Cmdr. Roullard, Asst. Naval Attaché. While there, they noted the large quantities of Japanese material that had been shipped to the USSR.

Because of the uncertainty surrounding the departure of the unit, a brief Christmas party was held jointly by the officers and men on 23 December 1945. In addition to songs and skits, another special goose dinner was prepared. Very late in the evening a group trudged through the snow over to the Uskievich dacha, marched in, and stood around their bed singing Christmas carols. Fortunately Byron Uskievich and his Russian wife Margarita rose to the occasion and stood on their bed in long night gowns and joined in the merriment.

Comfort projects.

It was up to each man to provide whatever comforts he could make with his own hands. Tables, book cases, clothes racks, and chests were in short supply. Every packing crate, for example, was used to fashion a useful piece of furniture. It was a challenge to see what could be made with the metal straps, nails and staples available on the crates. The traditional approach within the Navy for obtaining additional materials, known as "scrounging," was not possible in such a bleak and barren environment.

The most successful engineering triumph was the construction of a shower. A large, old samovar-like device was used to heat water with a wood fire outside the rear of the main building, which was situated on a bank. A pump raised the water, delivered by a tanker truck when available, to another tank under the roof. Gravity carried the water down to a small room at the basement level. The shower head consisted of a punctured tin can. The operation was limited by the delivery of water, the restricted times at which the pump could be operated, and the limited amount of wood available. Nevertheless, everyone had an opportunity to get several showers in the following months if they chose to endure the shock of cold air when the warm water flow stopped. Based on the squeals of delight, the cleaning women from the village enjoyed a shower for the first time in their lives.

The Russian engineer who supervised the plumbing for the shower became friends with Lt. Starke who showed his appreciation by giving the engineer a bottle of Spanish Brandy. The engineer in turn shared a bottle of blackberry wine that he had presumably made himself. Before the group departed, the engineer gave Starke a Russian-style dagger that he had made while serving as an aircraft mechanic. The double-edged blade was made from a section of a wing brace, the handle guard was fabricated from a billet of aluminum, and the handle was composed of several layers of different materials on a central shaft—kabob style. The memento reminds him of the personal friendships that developed in spite of the political, military and cultural differences.

Another project of benefit to the men was an indoor toilet in the barracks. The digging of a septic tank turned out to be a major enterprise. About 30 young Soviets were sent over to do the digging, and the result was a hole of a size that would serve the entire village! Again, getting water delivered on a regular schedule was the only deterrent to the use of the toilet. In general, the use of septic tanks was a rarity in the region. The outhouse was also improved slightly by providing a seat at a suitable height made from the wooden typewriter shipping boxes. The Russians still squatted over the box on the grounds that it was more sanitary.

One of the dachas, kindly made available to Lt(jg) Uskievich and his wife on her arrival (see below), had an indoor toilet, but the water supply line was above ground. The pipe quickly froze, and the facilities in the administration building had to be used. The walk in the frigid temperatures and snow required a major outfitting. For this purpose, Mrs. Uskievich was provided with a "tailored" set of Navy "long johns" to make the short excursion.

Missing mail.

It is vital for the morale of any military unit to be able to send and receive mail. When the last plane of the fourth flight took off on 4 October, it was the last opportunity to send mail directly home. On 8 November a courier arrived from Vladivostok, so the accumulated mail was transferred to him for delivery via diplomatic channels through Moscow. Some official mail was received by a second courier on 12 December, but it was all dated prior to the departure of the last part of the unit from the U.S. on 16 September! On 31 December, a member of the unit received a cable that the mail sent via Moscow had not been received as of 26 December 1945. Lt (jg) Nelson was expecting the birth of his second child and was obviously concerned when neither letters nor telegrams arrived. In desperation, he got the CO's permission to use Navy radio channels. After a series of relays, he learned that he had a second son Robert born on 27 September 1945. The Soviet authorities were no doubt swamped trying to translate the regular mail, but the failure to deliver mail in a diplomatic pouch needs explanation. Whatever became of the mail sent to the Fleet Post Office Box #1168, San Francisco, CA, would make an interesting investigation.

7. WEATHER CENTRAL OPERATIONAL

Purpose.

THE PRINCIPAL GOAL of a weather central is to collect all pertinent meteorological data within a given region, prepare weather charts, and finally prepare specific forecasts for a wide variety of operations. For example, the height of waves, swells, and surf was critical to landing craft; the direction and force of winds aloft were required by high-altitude bombers; aircraft carriers needed to know where to rendezvous in relatively good weather with aircraft that may have been in the air 8–10 hours; minesweepers had weather constraints for safety of operation; submarines preferred to surface under thunderstorms for protection; rescue helicopters (YNS or R4) were particularly limited by gusty winds; and there was a host of other weather-dependent naval maneuvers by the fleet. (One need not be reminded of the use of foul weather as a screen for the Japanese sneak attack on Pearl Harbor.) In brief, the weather central was the receiving center for all weather information that could be integrated with the sophisticated data collected at the center to produce a useful weather forecast. The analysis of these data was then distributed to the units engaged in operations that would be seriously affected by the weather. The distribution was accomplished by encrypted messages and charts sent out by radio and radio teletype. Communication was, therefore, a very important and essential part of the successful operation of a weather central. The radio "shack" was just that (Fig. 36). In spite of the primitive conditions, the equipment was maintained and operated in a very dependable way. A summary of the weather actually observed is given in Appendix B.

Data sources.

The Khabarovsk Weather Central was to be supplied with current weather data from Soviet weather stations east of 110°E longitude. To use these data, it was first necessary to get the location of the stations; their altitude, so the pressure

Fig. 36. Radioshack for Fleet Weather Central, Khabarovsk. Red army barracks outside the village Knaz Volkonka are in the background.

could be corrected to sea level; proximity to water, to evaluate the humidity; the topography, to determine if the wind was constrained in a valley, for example; and their schedule of observations. There was a haggle over each issue. The Soviets used a standard chart with a scale of 1:10,000,000, that is 1 inch = 160 miles, for plotting their weather data. According to Lt. Pede, no charts of a smaller scale were available. Fortunately, the Weather Central had a supply of meteorological charts (No. 5555, polar stereographic projection, from the Hydrographic Office, dated September, 1938) for that region of the Western Northern Pacific Ocean at a scale of about 1:15,000,000 (1 inch = 240 miles). The position of approximately 72 weather stations in 1938 were shown east of 110° E longitude within the USSR. Even so, the number of stations east of 110° longitude actually used by or for which data were available to the Soviets in drawing their own "synoptic maps" in the Central Institute of Prognosis in Moscow was about 20. The first weather map prepared by the Soviets available in the US Weather Bureau is for 31 December 1945 (Fig. 37). Apparently the weather maps drawn by the Soviets during hostilities (1937–1945) are still classified in the opin-

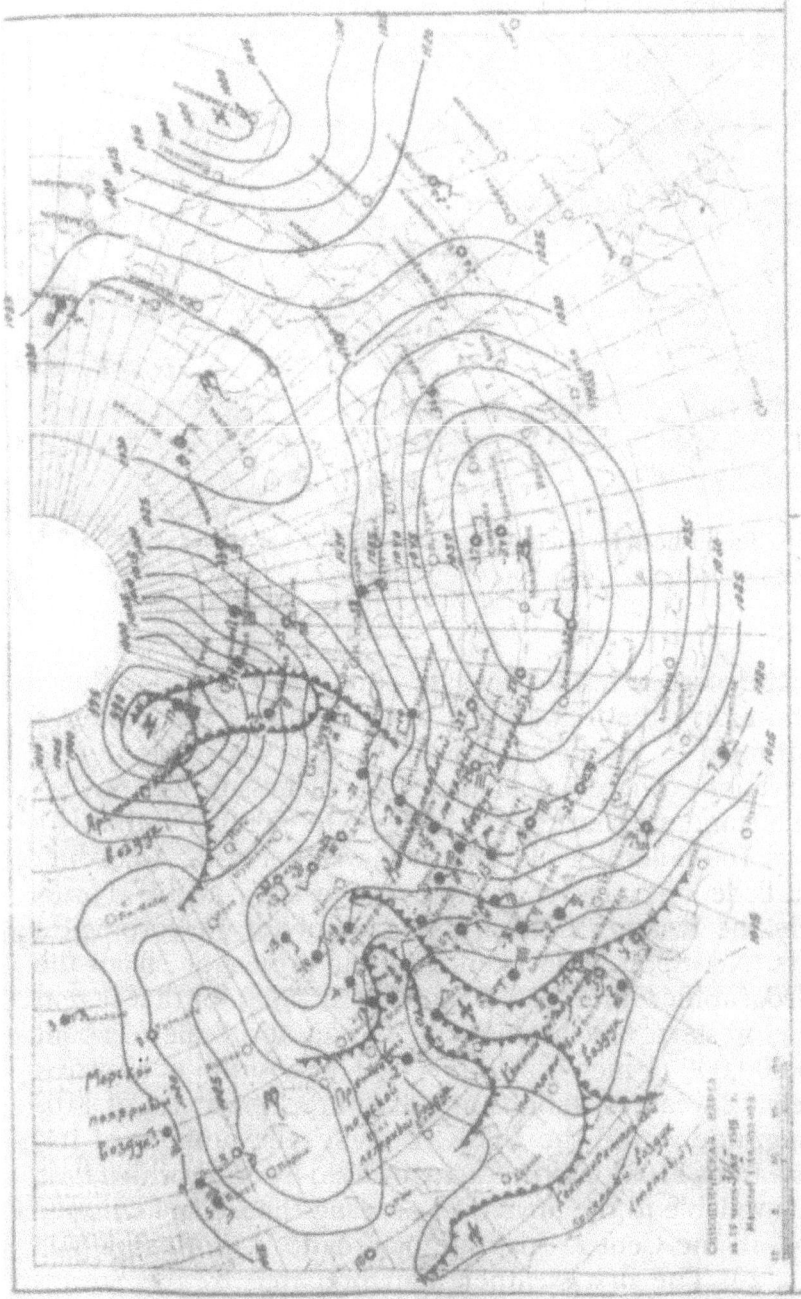

Fig. 37. "Synoptic map" of the Central Institute of Prognosis, Hydrometeorological Service of the USSR, for 1 January 1946, prepared at 1900 hours on 31 December 1945. In library of the National Oceanic and Atmospheric Administration, Silver Spring, MD.

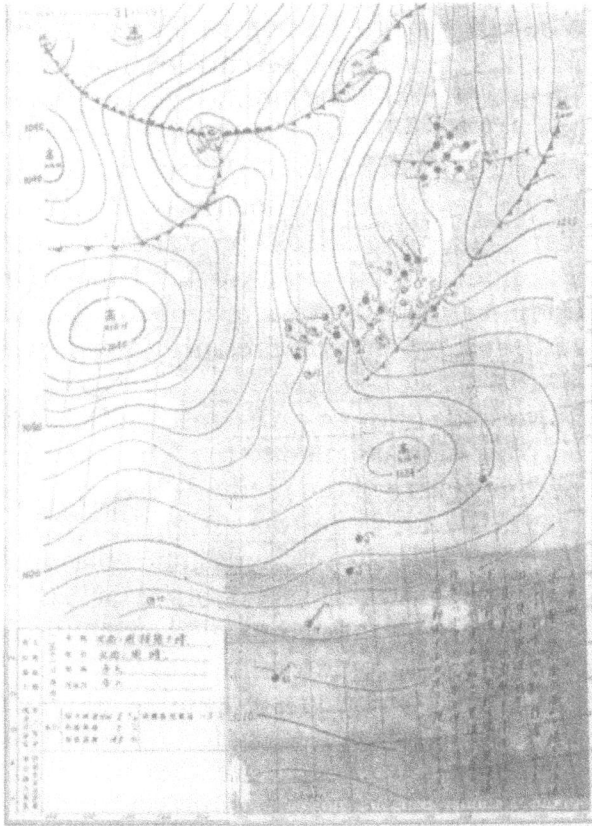

Fig. 38. Weather map drawn by the Japanese for 0600 on 31 December 1945 in the Central Meteorological Observatory of Japan in Tokyo. In library of the National Oceanic and Atmospheric Administration, Silver Spring, MD.

ion of a Russian meteorologist. (There is also a gap from 1937 to 1958 in the Soviet weather records held by the National Meteorological Library and Archive in Great Britain.) The comparable weather map for that date from the Central Meteorological Observatory of Japan in Tokyo is given in Fig. 38.

   The time of observations seemed to be a mysterious concept to the Soviets. It was definitely stated on one occasion that all observations were taken on mean solar time for the precise longitude of the station. On another occasion it was stated that they were governed by Moscow time. In fact, observations were made on a casual basis, and were worthless for synoptic map preparation where all observations in a region had to be taken at the same time. Furthermore, for the data to be of value they should be current, not six hours to a day late. The production of a synoptic weather map is an old idea (Brandes, 1826); however, having the data on a timely basis came about with the use of the telegraph. The Chief Signal Officer of the U. S. Army as early as 1871 (War Department Circular) had reports from 55 stations in the U.S. at three exact times daily in order to warn of approaching storms. In addition, it was never ascertained which Soviet stations reduced their barometric pressure to sea level, so the data were always suspect.

To resolve some of these problems, the CO asked that a Soviet meteorologist be assigned to the Khabarovsk Weather Central to become aware of the methods used by U.S. meteorologists. Eventually Lt. Pede was assigned, but he showed little interest in learning the methods and refused several invitations to visit the Central. To try another tack, the CO suggested that one of the unit's aerologists work in the Soviet weather office in Khabarovsk. Unfortunately, the Soviet office was being moved to new quarters and would not be operational for several months according to Lt. Pede. That excuse seemed to terminate any effective weather liaison. The Soviets claimed they had no weather stations on Sakhalin, Korea, and only two in the Kurile Islands. Reports from Mongolia and the Yakutsk region were very sporadic. The CO was assured that there were no weather stations in Manchuria even though flights were made daily into the region. On the other hand, the Soviets requested U.S. reports from the Aleutians, Alaska, Guam, Iwo Jima, Okinawa, China, Philippines and the U.S. weather ships in the Pacific: Lt. Pede was promptly given a copy of the Pacific weather broadcast schedule.

In spite of the continuing debate about Soviet data, cipher restrictions, communication delays, and general lack of cooperation, the Khabarovsk Weather Central established six- hourly transmissions of the local weather observations on 20 September 1945. It was equipped as a Class A weather station with regular six-hourly balloon soundings to ascertain the wind directions aloft and two radiosonde launches per day from which the temperatures, humidity, and pressures were obtained with height. From a study of the cloud formations, continuous record of surface temperature, pressure, and humidity changes, precipitation, and other details, it was possible for an experienced meteorologist to construct a reasonable representation of the weather map for a wide region. (A summary of the weather observed in the Khabarovsk area during the period of operation is given in Appendix B.) This type of single-station weather forecasting was essential at sea where the data from other ships had to be delayed in transmission to avoid revealing their position and possible attack by submarines. Nevertheless, the acquisition of data from Soviet weather stations would have enhanced greatly the accuracy of a forecast from the Weather Central.

Watch set.

On 12 October a conference of all of the aerological officers was held, and the CO stated that an all-out effort was to be made to prepare the first weather bulletin for broadcast on 15 October 1945. A log was to be kept of the actual material received from the Soviets to determine just what could be expected and used in the transmission. The office routine was planned and the duties of each officer outlined in writing. The watch was set. On 15 October, duplex radio teletype communications with Guam[50] were established and the first weather bulletin was published at 0005 GCT (Greenwich Civil Time).

Although an advance party of U.S. troops initially landed on the main island of Japan on 28 August 1945, before the Peace Treaty was signed on 2 September 1945, landings in force were carried out on 30 August 1945 by the 11th Airborne Division at the Atsugi Airport near Yokohama and by the 4th Marines of the VI Marine Division at the Yokosuka Naval Base on Tokyo Bay, both on Honshu Island (Williams,

---

[50] Fifty years later I learned that one of my eight classmates in the Navy Meteorological Program at the University of Chicago, Lt. George J. Haltiner, was the Senior Forecaster at the Fleet Weather Central in Guam, Headquarters of CINC-PAC. Cmdr. John Cory was CO of that Weather Central and Lt. Cmdr. William J. Kotsch (later Head of the Naval Weather Service in 1970 and promoted to Rear Admiral in 1971) was the Executive Officer. Kotsch and Haltiner had the responsibility of providing to Adm. Nimitz's staff aerologist, Capt. Anthony Davis, daily weather and sea forecasts for the seas surrounding Japan. (Major Bristor had served in Guam in the Army Air Forces Weather Office prior to his assignment as liaison officer to the MOKO Unit.)

1960, p. 551). Knowledge of the weather remained a critical factor because the landings had been postponed two days by inclement weather. It was well understood that the bulk of the Japanese Army was still undefeated and were in a position to resist any premature attempts at occupation. The official declaration that hostilities were terminated was not issued until 31 December 1946 by Presidential Proclamation—only two days before the MOKO group left Siberia.

Requests were renewed for getting the exact times when weather data were to be received from the Soviets. To cut down on the time delays, it was suggested that the teletype communications received at the Soviet station in Khabarovsk be sent simultaneously to the MOKO teletype. Lt. Pede said that would have to be discussed with his superiors. Locations of Soviet Stations were still missing, and as an incentive for a fair exchange, the grid for U.S. weather stations in the Pacific and China were given to Lt. Pede. Another difficulty arose because data were not received consistently from the same stations. Answers to the many questions posed were sought by telephone, teletype, and letter. A list of questions was eventually handed directly to Lt. Pede, but replies were never received. Even the request for simple climatological data was denied. The only bright spot during these frustrating times was that the CO of the Marianas had received the Khabarovsk Weather Central bulletins and found them satisfactory.

The forecasting of sea, swell, and surf conditions was left to the aerological staff in Guam. The theory and methods had been worked out by Sverdrup and Munk (1943; 1947) for the invasion of North Africa and tested by Crowell (1946) for two months prior to the Normandy Invasion, D-Day, 6 June 1944. No doubt the fiasco that evolved from the heavy swells, high waves, strong surf, and gusty winds at the time of the amphibious assault on Sicily on 10 July 1943 served as a wake-up call. The troops in the landing craft were violently sick, attack transports had difficulty off loading vehicles, parachutists were scattered or lost at sea, aircraft formations were shattered, and supporting dive bomber attacks delayed. As a result, the Hydrographic Office issued a confidential manual in November 1944, giving the principles for forecasting breakers and surf (Anonymous, 1944). None of the aerologists on the MOKO expedition was aware of these classified developments, but members of the staff in the Fleet Weather Central in Guam were knowledgeable about the critical need for accurate wind speed and direction to make such forecasts. With those data, wave height and wave period were forecast with due regard for coastal influences (e.g., wave break point and height with shallowing water, tidal effects, refraction of waves) and the moving areas of swell generation.[51] Equally important was the rate of decay of wind waves after the wind moderated or ceased because of the need for at least a three-day forecast. If the wind direction was accurate within 22.5° and the wind speed within one Beaufort force (a variable scale of wind force increasing in range with amount) the wave height was forecast with a maximum error of 1 foot for waves up to 5 feet and 2 feet for waves above 5 feet in height in the Normandy Invasion (Crowell, 1946).

Petropavlovsk connection.

On 7 October 1945, the Commander in Chief of the Pacific Fleet ordered the CO to send an officer to Petropavlovsk (Fig. 1) to see why that weather central was not in operation. The next day the CO requested permission from the Naval Mission in Moscow to send Major Kodis, an experienced weather forecaster who had served with the Army Air Forces group stationed in Poltava, Ukraine (see Postscript). About the same time Major Vishvarka was asked to get the necessary travel permits from the Soviet authorities in Moscow. In the meantime, a cable was received from Petropavlovsk asking for the frequencies to be used for the Khabarovsk-Petropavlovsk circuit. Apparently they were not to be granted permission to operate until the frequencies were received and checked by

---

[51] Decaying waves that have outraced their generating storm are referred to as swells. They become more uniform, regular in period, and have relatively long, flat crests. Their heights are generally small compared to their wave lengths. In the region between 40° and 50° South latitude, "the roaring 40's," of the South Atlantic Ocean, for example, the carrier USS Mission Bay (CVE 59) on which the author served incurred great risk in launching and recovering aircraft. The great swells, developing as a result of the unrestricted fetch of previous winds, prohibited launching even though the day was clear and the prevailing wind nil. The risks were in catapulting an aircraft directly into the crest of a heavy swell and the instability of a pitching landing platform.

the local authorities. The cable indicated that they had been
ready for operation since 20 September 1945! Permission to
visit the U.S. Navy Weather Station at Petropavlovsk was
never received. Radio communication was established about
9 P.M. 14 October 1945 with Petropavlovsk without further
discussion, and their weather report was included in the
Khabarovsk broadcast to Guam the following day.

The establishment of the Petropavlosk station (code
name TAMA) was an heroic effort. The group of eight officers
and 24 men under the command of Cmdr. C. J. McGregor
was literally dumped on the beach on 6 September 1945 and
expected to live in tents. The Navy Sea Bees realizing their
plight set up Quonset huts (Fig. 39) they brought after a
quick return trip to Attu, Alaska, on 18 September 1945.
Needless to say, the Soviets were equally quick to surround
the compound with a barbed-wire fence. The Quonset huts,
well insulated and fully furnished, were used for work, sleep,
meals, and recreation. Hot and cold running water, an electrical power system, and oil heaters resulted in a self-
contained and independent base (March, 1988). The Petropavlovsk group apparently received the same "hospitality" as
was experienced by the MOKO Unit.

Equipment installations.

The thermoscreen in which the weather recording instruments are housed was placed in front of the main administration building (Fig. 40). An anemometer, recording wind
speed, and a wind direction vane and transmitter were
mounted on the roof of the main building (right side of Fig.
40). Cloud heights were estimated during the day, however,
a vertical beam of light was used with a clinometer and a
known base line to measure the ceiling at night.

Balloon soundings were essential to record the winds
aloft. At night, a small candle was hung under the balloon
for visibility. A telescopic theodolite was used to measure the
elevation and azimuth of the balloon. A theodolite is essentially a surveying instrument and can be used for ground
mapping. When the uncooperative attitude of the Soviets became apparent, mapping of the underground storage facili-

Fig. 39. Quonset huts at Petropavlovsk, Siberia, used by U.S. Naval Weather Central. Photograph taken by U.S. Navy in October 1945.

Fig. 40. Thermoscreen, which protects the recording weather instruments, was located near the Main building.

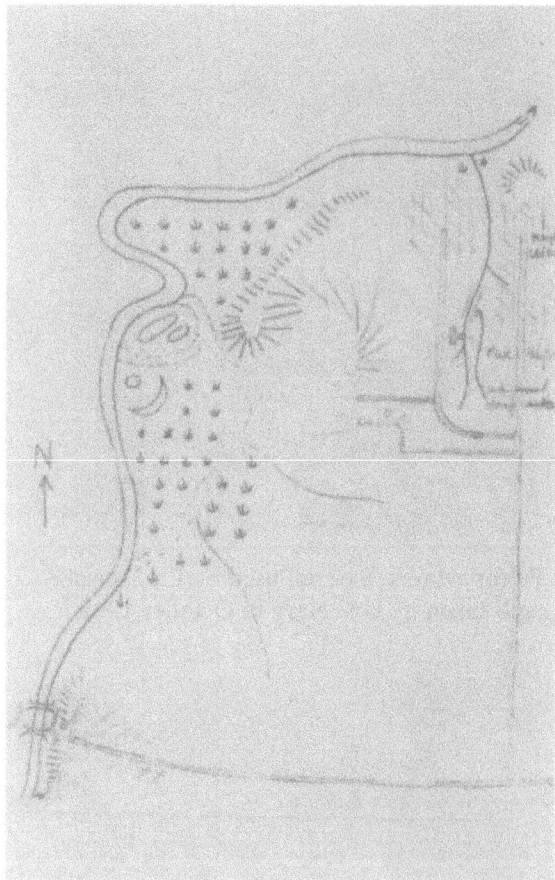

Fig. 41. Sketch map of MOKO Base near Village of Knaz Volkonka. Note general location of underground storage vaults and double bridge across Sita River. The by-pass bridge was constructed because the logs in the roadbed of the original bridge had rotted. The logs were laid parallel to the roadbed, and it was impossible to avoid getting the wheels stuck in the slots where the logs were missing. It is not known why planks were not laid across the roadbed on the many bridges crossing small streams.

ties along the Sita River (see Fig. 41) was undertaken personally, without the knowledge of anyone else in the unit.[52] By laying out a different base line each day for the regular balloon soundings and making back sights, it was possible to map the location of any distant object in view through triangulation. (To avoid discovery, the map produced had to be destroyed before leaving the USSR, but not before memorizing all the coordinates!)

---

[52] The seeds had already been planted in the my mind that someday we would end up fighting the Soviet System. Their primitive methods and brute-force approach to problems in that severe climate might be a severe challenge even for sophisticated, technologically superior weapons—and troops—not fully tested under such extreme conditions. The fatalistic resolve of the Soviets was a force that demanded the serious attention of any opponent. The risk of obtaining use-

On request, a chart showing the location of the Soviet stations reporting radiosonde and pilot balloon soundings was provided by Lt. Pede. Thirteen stations reported both types of data, and an additional six stations reported pilot balloon soundings only. Lt. Pede admitted that the list was incomplete and he would try to get a complete list. The complete list never came. He also said that the pilot balloon soundings only went up for a short distance even on a clear day.

The radiosonde was launched with a substantially larger balloon. Recording of its transmissions was done in the radio shack where technicians could monitor reception. The Soviets did not seem to be aware that data from a radiosonde could be checked by flying an aerograph on an airplane. The Weather Central had such instruments, but permission to make the flights had to come from Moscow. Permission never came. Because of the limited gas supply, eventually only one radiosonde per day could be launched. The Soviets produced their hydrogen by reacting ferro-silicon (FeSi) with caustic soda (NaOH). The generators had to be stored in a heated room to prevent the sludge[53] from freezing within the cylinders. They did not use the reaction of hydrochloric acid (HCl) with aluminum (Al) because of the high speed of reaction.

Generator housing.

Because of the extreme cold, it was difficult to start up and maintain the two diesel-powered, electrical generators. Even though snow fall was light in the area the CO ordered that a suitable building be constructed. Four enlisted men were assigned to me and two woodsmen from the local village were employed. The rugged but small woodsmen were a Uzbek and a Tartar. Neither spoke the other's language and neither understood Russian. A simple method of communication was developed whereby a tool was held up and alternatively each would assign a name. The noun was also used as a verb, but

---

ful military information appeared to be justified. After all, the U.S. Navy had not sent two trained intelligence officers along just to be translators.

[53] The reaction the Soviets may have used to produce the hydrogen gas is most likely: $2NaOH + FeSi + 3H_2O = 3H_2 + Fe(OH)_2 + Na_2SiO_3$. The brown sludge produced is probably composed of iron hydroxide, sodium silicate (one of the water soluble silicates), and excess water.

these highly skilled woodsmen needed very little direction. Post holes were turned out with an auger[54] and logs erected to support the walls. The cross pieces needed to support a roof without center support were made of overlapped boards pegged together. The shape was patterned after a sailing boom, thin on the ends and thick in the middle.

It was a joy to see the accuracy and speed of the woods-men in squaring a log. The final products seemed as if they had been machine formed, even though only an adze and axe had been used. The wood rang like steel as it was struck—the temperatures were -20 to $-25°F$. In only four days the frame was completed. The siding and roofing was made of sheets of Japanese plywood recently captured by the Soviets and transported in from Manchuria. Saw dust filled the space between the inner and outer plywood sheets for insulation. The building completed at the end of the seven-teenth day is shown in Fig. 42. The tubes protruding from the wall are the exhaust pipes for the two diesel generators.

The entrance door was on the far side and patterned after a barn door. Hinges were made available, but no latch was supplied. To keep the door closed, a pulley was rigged on top of the door frame and fitted with a rope and weight. In this way, the door was automatically closed by the weight. Within three days the concept was applied by the Soviets to the main door of the hotel in town. The physics was correct, but the weight was gigantic and it took all of one's strength to open the door to the hotel! The metal handles on the door to the hotel looked very much like the large handles used on coffins. We were told that there was only one multipurpose handle built in the USSR—the best.

The Uzbek and Tartar were very responsive and dedi-cated to the work. On several occasions they were invited to the author's quarters for some vodka and to see the *Life Maga-zine,* two copies of which had been smuggled in within the

---

[54] Augers had been brought in on the fourth flight for this purpose. The Soviet method was to dig a ramped trench and stand the log in the corner at the deep end, using the ramp as an aid to erecting the log. Because of the great interest in the hand-operated auger, one was given to a Soviet helper. In the following days, many holes were noticed on the parade field where he had been demonstrating its use.

Fig. 42. Building erected to protect two diesel-powered electrical generators. Note antenna tower on right. Red army barracks are in background. Roof of radioshack is just visible at left of generator building.

skin and lining of a plane on the fourth flight. (The vodka[55] was part of the daily ration allotted to each person on the base by the Soviets. The nondrinkers hoarded the ration for later barter. A bottle was worth almost a month's salary for a worker!) The response to the magazine was most illuminating, not only by the careful scrutiny of every advertisement but also the pictures of the U.S. countryside. Special note was made of the advertisements for ladies' lingerie, but the models were considered too skinny, if the sign language was correctly interpreted. (Most Russian women had the strength to bend a tall American into the shape of a pretzel with little effort.) The risk of seeing the magazines was great, but it

---

[55] Vodka is usually provided in 80 or 100 proof, that is, about 40 percent or 50 percent by volume of alcohol. The most pleasing volume of alcohol was determined by D. I. Mendeleyev (1865), of chemical periodic table fame, to be 40 percent . The reaction of water with alcohol is exothermic, and the mouth feels hot and dehydrated if the volume of alcohol exceeds 40 percent . On the basis of the observed freezing temperature, the vodka issued was probably nearer 100 proof, if only to prevent breakage of the containers during shipping across Siberia.

Fig. 43. The cover of the first issue of the illustrated publication "Amerika" (in Russian) distributed by the U.S. Bureau of Information about August 1944. The photograph of troops landing seemed to foretell the ultimate application of the weather centrals' contribution to the war effort.

seemed to be worth the risk to them. One of the woodsmen got the message across that the next place for them was Sakhalin, so they might just as well enjoy whatever pleasures were available.

A magazine called *Amerika* was published by the U.S. Department of State in Russian beginning in August 1944 for distribution in the Soviet Union (Fig. 43). In exchange, the Soviets distributed the magazine *Soviet Life* in English in the U.S.A. Both appeared to be patterned after the highly popular *Life Magazine*. Copies of issues Nos. 1 and 3 of *Amerika* were later obtained at the U.S. Consulate in Vladivostok, but none were seen in the hands of Soviet citizens in Khabarovsk.[56] A

---

[56] Publication was suspended for several years after 1953. The early issues obtained have been deposited in the Library of the U.S. Information Agency, Washington, DC.

copy of issue No. 2 is in the hands of Lt. Starke. Both the Soviet and American editions were filled with high quality photographs exhibiting the most favorable aspects of each country.

Emergencies.

Prior to the construction of the generator building, a fire occurred in one of the generators at about 2 A.M. on 5 October. A spark had ignited an open oil pan under the generator. The fire was put out by the unit's personnel in 25 minutes using sand and a carbon dioxide extinguisher. The Soviet fire apparatus arrived from Knaz Volkonka some eleven minutes after they were notified. The local fire department had no apparatus or materials to fight an oil fire. Fortunately, the damage was minor and did not affect the operation of the second standby generator.

On 12 October, Lt. Cmdr. Outcault was stricken with appendicitis. Dr. Revere thought that the operation could be performed more safely in a hospital. He was operated on successfully at the Soviet Army Hospital #301 in Khabarovsk with Dr. Revere observing. Visitors were asked to put on used white gowns that were hanging in an anteroom. We presumed that the gowns were helpful in limiting the transmittal of any diseases outside the hospital, but one wondered about the transmission of infectious diseases between patients. It was observed that the patient's temperature was taken under the armpit, not orally nor rectally. This procedure avoided the necessity of sterilization after use.

While in the hospital, Lt. Cmdr. Outcault was surprised at the large number of soldiers who had come from Manchuria.[57] They had the strange notion that the Americans were trying to surround them with troops in China, Philippines, Alaska, and now Japan. It was difficult to convey to them the vast size of the USSR compared to the U.S., our total lack of interest in acquiring more territory, and our strong resolve

---

[57] The Soviet losses in the Manchurian campaign were 8,000 killed and 24,000 wounded (Glantz, 1983, p. 219). The Japanese lost more than ten times that number in killed (83,737) and many prisoners taken (594,000) died in the slave labor camps in the following years (Weintraub, 1995, p. 95.)

to maintain friendly relations with the USSR. They were particularly proud of the notion that the USSR conquered the Japanese in one week[58] while the U.S. had spent years of fighting without success. Pointing out the large amount of Lend-Lease materials sent by the U.S. to aid the USSR had little impact. Again the jeep and Douglas airplane were claimed as products of the Soviets. No doubt Lt. Cmdr. Outcault was glad to be discharged from such a hostile and misinformed environment.

Another type of medical emergency arose for Lt(jg) Nelson. He vividly recalls having the local dentist install an inlay. The dental tools were "sterilized" by passing them through the flame of a Bunsen burner. The drill was mechanically operated with a foot treadle. (A similar drill was used by Lt. Starke's father early in his dental career in the U.S. when he made "house calls" in a buggy to his rural patients in the 1920s.) In spite of the primitive methods, Nelson said the inlay lasted longer than any of the others he had.

The unit was most fortunate to have Dr. Revere to deal with medical emergencies. If memory serves, he had passed his certification as a neurosurgeon, which required another three years of specialization after a two-year internship. As a result of his training, he developed an unusual skill that many wish they could acquire. He would announce that he was going to sleep for twenty minutes. Almost immediately he was able to drop into a deep sleep, but at the end of the announced time he would suddenly sit up. With a rub of his eyes, he was on his feet, bright and energized for his duties.

Doctor Revere immediately gained the respect of the "cleaning women." One woman had injured her finger, and

---

[58] The spectacular victory of the Soviets over the Japanese Kwantung Army in Manchuria was primarily the result of secret, massive (1.58 million men) redeployments of forces tailored to specific objectives with due regard for the wide ranges of terrain. The three frontal attacks were focused and used inclement weather, timed phases of battle, and darkness to achieve complete surprise. (The frontal system passing over Manchuria on 9 August 1945 is clearly indicated on the weather map for the Northern Hemisphere in the files of the U. S. Weather Bureau. In the view of Glantz (1983), the Soviets can indeed be proud of their impressive, speedy, and successful campaign carried out with "flexibility and audacity." As Glantz put it, "The Manchurian operation qualified as a postgraduate exercise for Soviet forces, the culmination of a rigorous quality education in combat begun in western Russia in June 1941."

it became seriously infected. He was able to arrest the infection and save her finger. All of the women eventually came to the doctor for gauze and cotton to attend to their hygienic needs. On another occasion he was able to set the broken arm of a workman from the village. He paid close attention to the health of all the personnel and there is no recollection of any serious problems. One advantage of the incredibly low temperatures was the absence of the common cold.

The most serious emergency resulted from a confrontation between the guard and the CO. After a particularly trying day, the CO decided a walk at night *within* the compound might help relieve the stress, in spite of the cold. For reasons unknown to anyone, the guard was not at his usual post, but he was inside the compound. He demanded the password— a recognition device that had never been used in the past. The customary challenge is "Stoy, kto idyot?" (Halt who goes?). Dumbfounded by the incredulous demand, the CO was speechless and was promptly fired on by the guard as he was trained to do. Fortunately, the shot was presumably a warning shot, and more importantly, missed. A Russian-speaking officer of the unit immediately ran to the aid of the CO and demanded an explanation. None was forthcoming then nor the next day when the event was related to the Liaison Officer. Discussion in the unit resulted in two possible interpretations. Either the guard, required to stand 12-hour-guard shifts with only a piece of bread and frozen fish to eat, was suffering from hypothermia, or, there was an attempt to remove the CO.

The CO had incredible patience and was able to maintain a civil response to the frustrations generated by the Soviets. He was persistent and innovative in his attempts to deal with them. Perhaps the Soviets felt he was too much of a match for their style of negotiations. But it was clear, the unit had to be on guard for the next round of tactics interfering with the effective operation of the Weather Central. Suddenly everyone felt that they were really earning their extra-hazardous duty pay, which was about 45 percent over base as Uskievich recalls.

8. SUSTAINING THE BASE

Moscow.

IN LIGHT OF ALL THE DELAYS, obstacles, lack of response to questions, and inability to send coded messages in the USSR, the CO decided to send an officer to Moscow early in October to brief the Ambassador. The failure to supply adequate water, food, wood, and fuel oil was critical. The unit was entirely dependent on daily delivery from Khabarovsk some 23 miles away on roads that were not always trafficable in the winter. His translator at meetings with the Soviets, Lt(jg) Uskievich, was chosen to make the long journey to Moscow to describe the difficulties of sustaining the base. The choice was made primarily because Lt(jg) Uskievich had served previously in Moscow and was well acquainted with procedures. In addition, the CO no doubt considered the fact that he had married a Soviet who was confined to Moscow. She was a beautiful ballerina in the Bolshoi Ballet Company, but lost all status because of her marriage to an American. Having lost her position in the ballet, she was without a ration card, shunned for political reasons, and dependent on friends for survival. Not without considerable intrigue, he managed to bring her back to the MOKO Base.[59] General Yarmilov had sympathized with Uskievich prior to his journey to Moscow and offered to provide his dacha (a log cabin used for sum-

---

[59] Uskievich petitioned the U.S. Embassy for a "travel permit" for his wife Margarita to accompany him back to Khabarovsk. Because Siberia was a military restricted region, Soviet approval was required. The Embassy's efforts were fruitless, so Uskievich called on the Commanding General of the Military Region of Moscow headquartered at the infamous prison "Lubyanka." The General was completely sympathetic and advised that no formal approval was necessary provided that the Embassy certified that Margarita was indeed his wife. The General also provided the special permission for them to travel by air and even arranged the transportation. Admiral Maples, then U.S. Naval Attaché, warmly congratulated Uskievich on his unusual achievement without the help of the Embassy.

Every effort was made to celebrate Margarita's 21st birthday on 19 November in the American tradition. Supplies were negotiated to bake a cake. Candles were made by melting down some of the candles used in the night time balloon soundings and dipping string into the melt to form the appropriately sized decorative candles. It was indeed a joyous occasion, even if the singing was off key and the words not understood by the celebrant.

mer vacation lodging) on the compound as living quarters if he was able to bring her back. It was indeed a friendly gesture that was greatly appreciated by the couple on their later reunion.

It took Uskievich five days to get to Moscow with stopovers at Tikda (=Tygda), Chita, Irkutsk, Krasnoyarsk, Novosibirsk, Omsk, Sverdlovsk, and Kazan. There was no night flying, the plane was not pressurized nor oxygen provided, even though altitudes of 14,000 feet were reached, and legs of the alleged through flight were canceled. Accommodations overnight were minimal, even when available. Uskievich describes "board spring bunks," several to a room, or sleeping on the dining tables. Food was even a greater problem. On occasion he was able to eat with the crew, but it was the radioman's job to pick up all the scraps after the meal for in-flight snacks for the crew. He shared some of his American canned food with the crew in repayment. On inquiring why there was no hot plate on the plane, the pilot replied that the hotplates had been removed by the Soviets because "Why have a hotplate when there is nothing [food] to heat on the stove." When the plane was parked some distance from the "terminal," an open truck was used to transport the passengers to the plane, or, on a plane change, the baggage was transported by truck and the passengers walked. Uskievich's return flight with his wife was equally distressing, with the discomfort stretching over seven days.

Station Wagon Acquisition.

On 1 November, the Russian-speaking ship's clerk, noncommissioned Warrant Officer Klopovic (Fig. 44), was sent to Vladivostok to arrange for the CO to meet with Majors Dalinin and Mukanov of the Red Navy Weather Service in Vladivostok. A supplementary map of the reporting stations in the Vladivostok area was obtained. In exchange, they were to receive the weather schedules of the U.S., which were delivered through the U.S. Consul in Vladivostok in a few days. In addition, Klopovic was to return to Khabarovsk with a U.S.-made Ford station wagon (Fig. 45), kindly supplied by the U.S. Asst. Naval Attaché in Vladivostok, Cmdr. George D. Roullard. Arrangements were being made through official

Fig. 44. Non-commissioned Warrant Officer J. Klopovic (right) with Russian jeep driver Tolya, who almost lost his life at the hands of a drunk Soviet officer, and Ens. L. W. Bowden (left).

Fig. 45. Cmdr. Butow with Ford station wagon provided by the Asst. Naval Attaché in Vladivostok and delivered to Khabarovsk by Warrant Officer Klopovic.

channels for the passes and documents needed to get from Vladivostok to Khabarovsk.

The U.S. Consul General, Mr. O. Edmund Clubb, was informed "that there would be very many obstacles en route" that would present hardships to the driver attempting to make such a trip. For this reason it was thought prudent to transport the wagon by train to Khabarovsk, and a request was made to the local authorities. The reply was that the Far East Soviet Railroad was already "overtaxed" as a result of military operations. In the meantime Cmdr. Roullard asked the chief of the railroad station if a flat car could be provided, and he said that would "present no problem at all." When the appropriate time arrived for transporting the car, the chief of the railroad station was nowhere to be found. That ended that attempt to obtain transportation.

After visiting the local automobile inspector to make certain which passes would be required, the inspector recommended he see the Soviet Diplomatic Agent. Recognizing the usual run-around, Cmdr. Roullard played another card by informing the Soviet Diplomatic Agent that if he did not hear from him by evening of the next day, it would be assumed that they had no objection to Klopovic driving the car to Khabarovsk. On the evening of 3 November Klopovic and Chief Yoeman Grason, stationed at Vladivostok, set off.

The first stop was the check point 19 kilometers outside the city, the limit of travel for persons in Vladivostok without special papers. Klopovic flashed a handful of papers with a big red seal affixed to them, and the guard was suitably impressed and let him pass. The road between Vladivostok and Voroshilov was hard surfaced with many windings and sharp bends through the mountainous region. They were not challenged at the several check points and, therefore, did not stop. Two bad bridges were detoured by fording directly through the river or finding an ice-covered portion thick enough to support the car. A large number of guards were at the Voroshilov check point because the city was quarantined. On the grounds that they only wanted to drive around the outskirts of the city on their way to Khabarovsk and were on urgent business for the American Consulate, the sentry motioned him on. The next ten miles, in Klopovic's words, "was the closest thing to a tank obstacle course that we had ever ridden on." He arrived with a broken spring that had to be welded by a local mechanic. No doubt the weight of a large number of cases of Scotch obtained from the consulate contributed to the failure of the spring.

The trip was planned to take advantage of the night when they would not attract attention. North of Voroshilov the roads were in poor repair, the long bridges of questionable safety, road signs were absent, but the principal hazard was parked trucks *in* the road! Drivers would simply turn off their lights and go to sleep until daylight to resume their journey. From Bikin, a mining town housing Japanese prisoners, to Khabarovsk, the road improved, and the guards simply motioned the car through. The drive took exactly 17 hours, whereas the Moscow express train took 21 hours to cover the same ground. The station wagon was all the more appreciated after hearing of Klopovic's experiences.

Vladivostok.

In light of Klopovic's experiences, the CO immediately set off for Vladivostok by train[60] to work out the details for acquiring the Vladivostok collective weather broadcasts. He urged the Soviet officials to transmit regularly and accurately the weather data so critical to the operations around Japan. The negotiations were interrupted on 7 November in order to attend the 28th anniversary celebration of the October 25th Revolution. The change in date resulted from the shift of thirteen days in February, 1918, from the Julian calendar to the Gregorian calendar (Fig. 46). After a detailed explanation of the methods of weather analyses used at the Weather Central, a technical discussion on the problems of long range forecasting ensued. No invitation was issued for the CO to visit the Soviet weather station or its offices. The Soviets again stated that no weather information was being received from Manchuria, Soviet ships, or Soviet aircraft. The only substantive result of the conference was a schedule of the

---

[60] The train never exceeded 25 miles per hour and made stops every 20 to 30 minutes lasting 5 to 10 minutes. No heat was provided on the train. The trip of approximately 400 miles on the local train from Khabarovsk to Vladivostok took 24.5 hours.

Fig. 46. A page from a
simple desk calendar for
1945, illuminated with a
scene from the Revolu-
tion, dated November 7,
Wednesday. "28 years
after the magnificent October socialistic revolution."
"1943— Division of the
Red Army liberates Fastov
from the German occupa-
tion." The slogan on the
flag reads "All power to
the Soviets."

weather broadcasts from Vladivostok and a list of Soviet stations along the eastern Siberian coast. It also established the
fact that there was little cooperation between the Khabarovsk
Weather Central run by the Red Army and the Vladivostok
Weather Office run by the Soviet Navy (CO Report No. 27,
1 November 1945).

Logistical support.

The CO spent the 9th of November with the liaison officers
of the Assistant Naval Attaché at Vladivostok. The possibility of U.S. ships docking at Vladivostok with supplies for the
Weather Central was discussed. In addition, the prospects of
sending an officer from the MOKO Unit to Petropavlovsk

via a Soviet ship were investigated. While granting that both proposals were feasible, the procedure would have to be investigated.

Other avenues for supplying MOKO Base were being investigated. Supply by the Naval Air Transport Command was considered impractical. The use of the Marine Transport Squadron at Yokosuka, Japan, was another possibility. Two ships were to sail from Portland, OR, carrying previously authorized Lend-Lease supplies, so attempts were underway to get supplies and mail for MOKO aboard. The more direct approach was the most successful: U.S. Navy survivor's food from the stockpile at Murmansk was requested. Some 29 tons of food, purchased by the unit, arrived in Khabarovsk on 22 November and was transported to the base the next day. The Thanksgiving Day celebration took on a very practical meaning. Nevertheless, the problem of future logistical supply remained, and the CO requested information on plans for continued support from the Chief of Naval Operations.

Photographs of the personnel, buildings, and some equipment installations were taken on 22 November at the request of the CO by a photographer (Shura Vishkvarko?) provided by the Soviet Liaison Officer in that no cameras were allowed in the unit. The CO thought it appropriate to repay the photographer for her cooperation with some of the newly arrived supply of canned Vienna sausage and tuna. She met Birkett and Klopovic, who spoke Russian, at the door in a heavy quilted parka. The gift of such a large amount of food brought tears to her eyes. Because it is the Russian custom to return a gift of equal value, she realized that was not possible under the circumstances and excused herself for a few moments Birkett recalls that she returned in flimsy clothes in spite of the cold with the intent to repay their valuable gift with the only gift of value she had—herself. With profuse thanks for her gesture, the two officers departed immediately in good conscience.

Winterization.

As early as 5 September the CO expressed concern about the fact that winterization of the unit's quarters had not begun. After all, it was a summer recuperation camp. The issue was raised again on 22 September. Changes in the original plans for winterization, alteration, and repair had been made by the assigned Soviet engineers without the approval of the CO. An offer was made to lease the land and buildings in the name of the U.S. Government with funds already deposited in the Khabarovsk State Bank.[61] By this action winterization could be carried out without making continual demands on the Soviet military. Major Vishvarka did not think that arrangement was possible, but stated he would consult the proper authorities.

Discussions again on the subject at the end of September were not fruitful. The major did say on 9 October that all possible speed was being taken. The slow progress on winterization was brought up again on 12 October; so the Major sent a dispatch to Moscow requesting higher priority for the construction company to complete the work by 25 October. The deadline came and went without any action.

By 1 November the wood supply was exhausted, and no attempts were made by the Soviets to deliver more than a day's needs at a time. A wood cutting party was organized to alleviate the wood shortage. After a long lecture by the Major on how wood cutting was programmed with regard to where and how much, the wood cutting party was dismissed. More wood was suddenly delivered by the Red Army (Fig.47). "Winterization" became a personal enterprise by wearing as much winter gear as possible and drinking goryache chai (hot tea). Thanksgiving day was spent by the fireplace.

The wood supply incident raised the general issue again about keeping the buildings warm. Two more stokers were needed for the stoves in the men's barracks to keep a minimal amount of heat available for those men off duty. A small boiler produced hot water for a few radiators in the administrative building, but they were totally inadequate for the job. Major Vishvarka promised Cmdr. Butow that help was coming, but the men, recently discharged from the army, had not yet obtained "passports." If help was not forthcoming the CO was prepared to move the men into the administration building and move the officers into the three dachas.

---

[61] About $10,000 (~53,000 rubles at the offical exchange rate) had been deposited with the understanding the Bank would hold the funds until notified of its disposition.

Fig. 47. Woodcutters at work trying to provide fuel for the many stoves in the "summer" camp.

That argument seemed to have achieved the desired result, and the boilers and stoves were tended 24 hours a day. Commander Butow then asked for a list of the current civilian help and their duties. The list (Table 2) was provided for a total of 19 persons, but the three additional people promised (two stokers plus a girl to assist in the men's galley) were not included (CO Report No. 28, 1 November 1945).

Petropavlovsk visit.

A third attempt was made to visit the weather central at Petropavlovsk. The need for logistical support was critical to both Centrals and a common solution to the problem was preferred. The CO decided to try a direct commercial flight to Petropavlovsk, taking Lt. Worchel as liaison, to work out the arrangements for supply. Major Vishvarka learned of the trip and made reservations for them through the colonel in charge.

Arriving in time for an 8 A.M. departure on 4 December, the ticket office said the plane was to carry cargo and no passengers. The pilot loaded the plane and said he could actually take them. A very cordial colonel explained that they needed passports. Attempts to see the military chief of staff and the liaison officer failed because everyone was "unavailable." That afternoon Major Vishvarka was located who said only General Ivanov could issue the required passports. The CO showed the letter from Admiral Nimitz ordering him to visit the Petropavlovsk Station. The major said he would be glad to forward the letter to General Ivanov, but he had gone to Manchuria. The CO got the message and returned to the unit's base the following morning. The reason for the runaround became crystal clear when a dispatch from Admiral Nimitz, Commander in Chief, Pacific Fleet, arrived.

9. GENERAL IVANOV CLOSES BASES

The edict.

THE RED ARMY CHIEF OF STAFF requested on 5 December 1945 through the Moscow Naval Mission that Admiral Nimitz close the Khabarovsk and Petropavlovsk Weather Centrals by 15 December. General Ivanov, Commander of Soviet Armies in the Far East, said the weather centrals were unnecessary due both to the regular Soviet weather transmissions being supplied and to the cessation of hostilities. There was no mention of the extreme burden the Centrals had become for the local administrations in supplying food, fuel, and water; the unacceptableness of the transmission of coded messages by the units; nor the great difficulty the Soviets were having in delivering quality weather information on a routine schedule. There was also the feeling that the Soviets had secured what they wanted (i.e., Lend-Lease had been terminated 20 August 1945), so there was no need to fulfill their end of the bargain. In addition, the Soviets might have been concerned about the unit's personnel contaminating the local citizens with capitalistic ideas. Another angle was perceived that the Soviets wanted the equipment and could acquire the Central by forcing the personnel out. In view of the general suspicion the Soviets had for the weather centrals and foreigners in general, they may not have been willing to extend any potential risk even though no alternative subversive use could be deduced for the weather centrals.

The reasons for the run-around regarding the flight to Petropavlovsk, or any visits to or from there, became quite clear. The foot-dragging on winterization of the "summer" camp and the delays in delivery of supplies were no doubt the result of an early decision to prevent the units from operating.

Unit informed.

The Khabarovsk Unit received the news on 7 December and the officers were informed of the Ivanov edict. (The date was probably fortuitous, but it raised strong feelings among those who had been in service on Pearl Harbor Day.) Liberty was discontinued the following day and an inventory of all materials was undertaken. There were just eight days supply of diesel oil to run the electric generators—as if the local Soviets already knew when the base was to close. Food and water were critical and even the gasoline for the station wagon had been measured out for the appropriate period. All hands were assembled on 11 December to advise them that written orders had been received from the Chief of Naval Operations via the diplomatic pouch to Vladivostok to close the unit on 15 December. Preparations were to be made to return to the U.S. via ship from Vladivostok with all personnel as well as all aerological and weather equipment. Capt. Cumberledge said the original plan was to leave all equipment on departure, but because of the Soviet attitude, Admiral King, Chief of Naval Operations, instructed us to "to pack it all up and bring it back." At the same time, the local Soviets were informed the unit would cease operations on 15 December 1945 as requested. Instructions were requested from Guam for evacuating the 60 personnel and about 80 tons of gear. The Commander in Chief of the Pacific Fleet advised the CO to be in Vladivostok on 27 December. Because the opportunity to return via air was denied, the S.S. Edward J. Berwind had departed the U.S. and was expected to be in Vladivostok on 28 December.

Shutdown.

The last weather bulletin was transmitted at 11:30 on 15 December and the transmitter closed down at 13:58 as ordered. Major Vishvarka said that passenger, baggage, and freight cars would be provided. Furthermore, he agreed to provide 40 trucks and a number of men to help in the transfer. Because of the potential delay in arrival of the ship, permission was requested for the personnel to live in the railway cars on a side track until the ship arrived. The CO stressed the necessity of moving the unit as a group. These plans were relayed to all officers on 16 December with the admonition that extreme conditions of discomfort were anticipated and the officers were to bend every effort to provide for the men. All printed regulations and instruction books were to be burned.

Mayor's Visit.

The mayor of Khabarovsk, Mr. Sergiev, called unexpectedly on Captain Cumberledge around noon on 21 December. It was his first visit to the Weather Central. The captain invited him to lunch, and during the ensuing conversations asked the mayor how the Americans had conducted themselves while in the city. The mayor replied that "all those with whom he had talked were extremely impressed with the courteous and gentlemanly conduct of the Americans, and that deep down in their hearts they had acquired a strong feeling of friendship for us. He hoped that the people in Khabarovsk had made us feel at home here during our short stay." The Captain replied that the personnel appreciated their cordiality and that the Russians had gone out of their way to make our stay as comfortable and enjoyable as possible (CO Report No. 38, 21 December 1945). As you can see, the conversation was carried on at a very high diplomatic level of politeness.

Captain Cumberledge told the mayor of his intent to give the remainder of the unit's food supplies to the orphans of Khabarovsk. Approximately 17 tons was thought to be available after laying aside sufficient quantities to sustain the unit for ten days in case of unforeseen delays in our departure. The mayor was most appreciative of the gift, especially because the food was jointly owned by the officers and men. The food was to be transported in six Studebaker trucks to the local Soviet orphanages though arrangements by the Liaison Officer. As a parting gift the mayor gave Capt. Cumberledge a set of Siberian moose horns, which he still has to this day.

Packing up.

The schedule called for loading of equipment on the flat cars on 24 December, so packing was undertaken at a furious pace. One officer passed out from exhaustion as a result of the long hours and extreme temperatures. Installations were dismantled even though the day-time temperatures ranged

from $-38°F$ to a high of zero. Working at sub-zero temperatures took some adjustment. Either hypothermia set in or
plain stupidity took over as one struggled to take a nut off a
bolt. In frustration the mittens came off, and as soon as the
cold metal was touched, you and the metal were one. That
meant the cold bolt or wrench had to be warmed up in the
pocket to free your fingers. Much of the equipment was returned to the original packing cases, which had been adapted
to other purposes. Because the lumber, nails, matting, and
straw requested from the Soviets were slow in coming, those
items were taken, therefore, from the local buildings.

As luck would have it, approval of a city telephone line
between Knaz Volkonka and Khabarovsk was received on 18
December. It was learned that the Liaison Officer had held
up the telephone line, because he wanted all communications
to pass through him. No doubt he now felt secure in knowing that the base ceased operations three days before and
the phone might facilitate the departure. Another item requested for the men's recreation hall, which had been delayed, a piano, also arrived the same date. At least it was
available for the brief Christmas celebration. Christmas Day
was to be a working day except for a two-hour program to be
produced by the officers and men. Voluntary contributions
of candy, cigarettes, and toilet articles were distributed as
gifts to the Soviet civilian servants working on the base. The
older servants cried on hearing the Christmas carols sung by
the assembled group. A more disconcerting delay was the
lack of mail that had been sent via courier on 9 November
and had not yet arrived. In addition, supplies that had been
sent on 3 December via the S.S. Wallace R. Farrington from
Portland, OR, did not arrive.

Because of the shortage of food, a local custom was investigated. The Sita River freezes to a considerable depth,
leaving rivulets and pools of water at the bottom. The trick
is to find clear ice over a pool in which fish may be trapped.
By chopping out a hole in the ice, one can reach in to grab
or spear a fish of your choice! One boy was observed to chop
out even the dead fish in the ice. That was one fish story that
had a believable and tangible result. (Birkett wondered if
the absence of cats and dogs was related to the food shortage.)

Assembling a train.

On Christmas Day the Asst. Naval Attaché at Vladivostok advised the CO by telephone that an attack transport was being rerouted to Vladivostok to evacuate the unit on 27 December. Orders were issued, therefore, to commence evacuation. As expected, none of the transportation equipment was available on 26 December as promised. The liaison officer was "unavailable" but after repeated appeals, nine Studebaker trucks arrived. Last minute packing was completed—at the expense of the materials in the local buildings—and about half the gear was taken to a railway yard siding. No freight cars were available, so guards were posted at the equipment site. The drivers of the trucks refused to work after 9 P.M. By 10:30 A.M. the next day, all the equipment had been transported to the railway siding, but no railway cars were in sight.

It was decided that a more direct approach was needed if railway cars were to be obtained. That night several officers and men proceeded to push whatever cars in the yard could be found to the siding. The process was not without risk because it was the custom for the Soviet railroad guards to fire a burst from their submachine guns up and down the tracks from their stations to prevent the local citizens from stealing coal that had dropped off the cars on to the track. Four box cars were found; however, no flat cars were seen. Because of the cold and obvious signs of frostbite (e.g., white spots on the facial skin[62]), the group went into the railway station. There it was learned that a flat car had been unloaded about a half mile away. The station master even promised a switch engine after midnight, but it never arrived. Again, the group pushed the U.S.-made flat car to the siding. By 10 P.M. the next day all the gear was loaded into the railway cars, including the two three-ton generators. There was no room for the station wagon, so it was necessary to drive it down to Vladi-

---

[62] In contrast to the white spots were the red dots formed from frozen blood drawn by the penetration of ice needles blown by the wind. The white-outs produced by blowing "snow" resulted in disorientation, but the railway tracks at least gave one two choices of direction.

vostok. There still remained the problem of getting an engine to pull the train to Vladivostok.

The attack transport USS Starr (AKA67) docked at Vladivostok on 27 December as promised. It was agreed between Captain Withers of the Starr and the U.S. Asst. Naval Attaché in Vladivostok that the unit would be boarded at 11:00 P.M. on 29 December. From unofficial sources it was learned that it would take two days to get the freight to Vladivostok. In order to get an engine, it was necessary to take some unofficial steps. With suitable lubrication and the promise to each of a case of vodka—worth almost a year's wage—an engineer and fireman were "persuaded" to make the run. Two men were assigned each freight car to guard the contents of the cars. The Soviets would not assume any responsibility for the shipment.

Even though the Soviets were bent on *not* forwarding the equipment, they did provide a baggage car and a passenger car on a separate train two days after the promised date. The train departed at 11:00 P.M. on 28 December 1945 with all hands, except Ship's Clerk Klopovic who drove the station wagon. The unit was wished a friendly "Good-Bye" by the Chinese Consul. There were no Soviet representatives present except the Liaison Officer who appeared after he was assured the unit was really leaving. He supplied a statement showing that there were no outstanding obligations. Reverse Lend-Lease papers were signed for materials received. It took 24½ hours to reach Vladivostok after passing through innumerable check points. Klopovic in the station wagon had to pass through an even greater number of road check points, but succeeded by waving a letter in English signed by General Ivanov. The letter had nothing to do with travel permits, but the signature and the conversational skills of Klopovic resulted in safe passage.

The most impressive sight during the train ride to Vladivostok was the mountains of contraband the Soviets had taken out of Manchuria. There were piles of large machinery of every description left to rust in the weather alongside the tracks. There appeared to be no attempt to salvage the valuable tools. It seemed as though the intent was to strip Manchuria of any production facilities. There was no sorting to suggest that the metals were to be used for scrap. Even brass beds and steel bed springs were in the jumble. Presumably all usable materials had been spirited away (e.g., the Japanese plywood used in the generator shed at the weather central).

Whenever the train slowed down in anticipation of a station stop, the passengers would grab a container and run as fast as they could to the station office. The container was filled with hot water to make tea, and if the line was not too long, one made it back to the cars before the train left the station. The passenger car trainman had a small charcoal stove on which to boil a little water. They were generous in providing a hot drink to the unit members, but food was something one had to bargain for from the rare vendor along the station track.

Civilian or Military?

The Assistant Naval Attaché had been forewarned of the unit's arrival and arranged for truck transportation, not to the dock, but instead to the customs office! At one o'clock in the morning, each member of the unit had to have at least one bag inspected carefully. The Custom House staff was not able to search the entire 12 tons of baggage. The CO reminded the Soviets we were military allies, and protested the entire procedure. All personal letters, books, notebooks, etc., were examined by a Soviet translator. All personnel were searched even though the Red Navy admiral in Vladivostok recommended clearance without inspection. (It was later learned unofficially the customs agents received a dispatch from the Soviet Red Army Staff in Moscow demanding an explanation for the searching of the unit's baggage.) Without food or sleep, no one was in any humor to deal with the bureaucracy. At about 3:30 A.M. the unit was finally permitted to board the USS Starr. The ship's crew had been granted liberty, but the Soviets refused to let any of the unit's personnel off the ship.

In the meantime, the freight train got underway a little after midnight on 30 December and was expected to reach Vladivostok at 2 P.M. on 31 December. When that time arrived, it was learned that the freight was in Voroshilov[63] and could be expected at 11:00 P.M.[64] After consideration of the working habits of the Soviets, particularly at night, the CO ordered that the freight be unloaded at 6 A.M. on 1 January 1946. There was no report on what became of the engineer and fireman and the two cases of vodka, but they may have escaped punishment in the exuberance of the New Year's celebrations. In spite of the great efforts to retrieve the freight, virtually all of the MOKO equipment and material returned to the U.S. Naval Depot at Mare Island, CA, was later destroyed according to an eyewitness.

Over 100 ships at anchor were locked in the ice about 8 inches to one foot thick in the harbor of Vladivostok, described as an "ice-free port." The pack ice was sufficiently thick to support small vehicles. The crews, mostly women, walked to and from their ships on the ice. The ships were of minimal construction and had very few facilities for the crew. For example, a wooden outhouse hung over the side of the ship. The limited period of service planned for these ships probably accounted for the conservative use of materials—a concept also employed in the construction of the U.S. Liberty ships.

The smaller longshoremen and women wore goat skin coats 'liberated' from captured Japanese. It was said that hundreds of Japanese prisoners were being used as dock workers, contrary to the Geneva conventions. The removal of a few cans of food from the packing cases being unloaded appeared to be a common practice that was overlooked by

---

[63] Voroshilov, now called Ussurijsk, lies 40 miles north of Vladivostok. It was named after Marshal Voroshilov, a hero of the Civil War (1918–1921), who served as Commissar of Defense from 1927 to 1940. It was a terminal on the original Trans-Siberian Railroad that passed directly across Manchuria in 1908. The younger Soviet section via Khabarovsk, wholly within the USSR, joins the Manchurian section near Chita.

[64] An article in the *New York Times* for 21 March 1946 reported that the freight cars "actually got separated and lost on two separate occasions," quoting the CO, Captain Cumberledge. He expressed the view that this was "due primarily to inefficiency and mismanagement rather than any deliberate attempt to delay proceedings." That diplomatic stance skillfully covered up the attempts of the Soviets to retain the advanced aerological equipment used at the weather central.

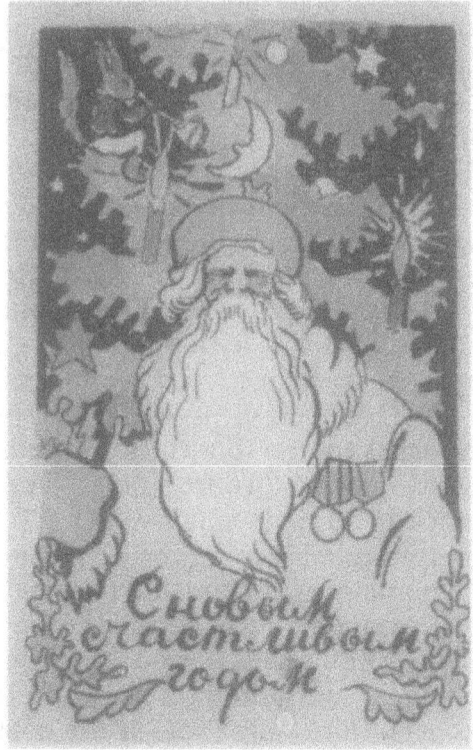

Fig. 48. A New Year's celebration card. Note the medals on St. Nicholas, a Russian patron saint. Candles and ornaments on an evergreen tree, although an ancient custom, emerged as early as the sixteenth century in Germany as the "Christmas" tree. The patron saint is referred to by the Russians as "Grandfather Frost" and the tree as the "New Year's tree."

the substantial number of guards. It was not determined if the guards shared in the loot.

New Year's celebration.

On the eve of the New Year, permission was granted for some of the officers of the unit to attend parties at the Soviet Armed Forces Club, the home of the Consul General of Vladivostok, and the U.S. Asst. Naval Attaché's home. Several instances were reported at the Soviet Armed Services Club in which Red Army officers "advised" the female dancing partners not to associate with Americans. One Red Army Officer soon discovered that some of the unit's officers could speak and understand Russian and offered profuse apologies for their advice to the ladies. On the other hand, New Year's cards expressing good wishes for the coming year (Fig. 48) were given to those attending the party. Most of the unit retired early because of the grueling conditions endured during the transfer from Khabarovsk to Vladivostok.

There are vague recollections of a visit to the U.S. consulate in Vladivostok. Soviet guards were posted at all corners of the compound. The old question of protection versus containment arose immediately. The consulate was highly restricted in his movements and apparently only then with permission from local authority. A visit to the Chinese Consulate revealed a similar situation. The consul expressed concern over the million Chinese that had disappeared into Siberia, and he had been denied permission to visit or speak to any of them. It was pointed out to members of the unit that there were few stone buildings in Vladivostok and the principal brick building was the barracks built by the U.S. during its occupation[65] in 1918–1920. Needless to say, the resentment was still evident.

---

[65] American troops arrived in Siberia on 16 August 1918 under the command of Major General William S. Graves. The mission was to assist the Czech Legion fighting its way eastward along the Trans-Siberian Railroad, discourage Japanese intervention and appropriation of parts of Siberia, and prevent the remote possibility of the Germans taking the munitions stored in the area. The orders were explicit not to become involved in the internal affairs of the USSR then engaged in a civil war (Graves, 1931). Whatever was done was interpreted as helping the other side, so the resentment of the American presence ran high. The troops were evacuated in March, 1920.

The Soviet Union was recognized by President Franklin D. Roosevelt in 1933. A courtesy call was made to Vladivostok by the U.S. Navy on 28 July–1 August, 1937. The cruiser Augusta (CA31) and four destroyers under the Commander in Chief of the Asiatic Fleet, Adm. Harry E. Yarnell, visited this vital port. It was alleged that no U.S. warships had since been in that port until the USS Starr arrived on 27 December 1945.

10. ESCAPE FROM THE SYSTEM

Homeward bound.

THE USS STARR sailed from Vladivostok at 3:45 P.M. on 2 January 1946.[66] Ice breakers and mine sweepers had already cleared a path to the open sea. The feeling of relief from escaping the Soviet system was indeed great. The happiness of the unit was tempered, however, by the heart-wrenching scene at the dock. It was not possible for Lt(jg) Uskievich to take his wife on board (Fig. 49). There was some discussion with the captain of the USS Starr about taking her and Irene Matusis on board, but Uskievich did not want to jeopardize the position of the wives of other Americans in the USSR that he knew personally. Even Major Vishvarka tried to be helpful and urged that Margarita stow away on the U.S. Navy Ship! The captain of the USS Starr, well aware of the food shortages, did provide a week's provisions for her to carry her through the seven-days journey to Moscow and to her anxiously waiting family. Mrs. O. Edmond Clubb, wife of the American Consul General at Vladivostok rendered superb support by sheltering Mrs. Uskievich at the consulate and in preparing her, then in the early stages of pregnancy, for the arduous trip to Moscow. As all waved good-bye to her, the NKVD took their positions behind her. Eventually she made her way to the U.S. but not without the help of the Secretary of State and others.[67]

---

[66] The Petropavlovsk Weather Central had already been evacuated by the USS Sarsi (ATF-111) about 19 December 1945.

[67] It was later learned that after the highest level of negotiations, Lt(jg) Uskievich's wife Margarita was given permission to leave the Soviet Union. Uskievich even made a direct appeal to Stalin! During a cocktail party celebrating the transfer of 21 mine sweepers from the US to the USSR, Uskievich met a Soviet general of the NKVD. The general could not understand the inability of the State Department to get Margarita an exit visa, and promised to look into the matter. Later, with the kind assistance of Senator Theodore Francis Greene (RI) and Secretary of State Dean Atchison, an exit visa was granted to Margarita and four other Russian wives of Americans. While the negotiations lasting many months were underway, their son Richard was born on 4 August 1946. Margarita appeared with her clearance papers only to learn that she had to have separate papers for

Fig. 49. Mrs. Margarita Uskievich (left) with Maria, maid for their dacha, and Lt(jg) Byron Uskievich.

The principal effort on shipboard was to prepare a history of the expedition, but limited solely to the weather operations by order of the CO. All the facts and figures that had been memorized were jotted down in the safety of a U.S. warship. The attack transport was to stop at Sasebo, Japan, to pick up units of the U.S. Marine Corps being returned home for discharge or duty rotation.

---

their son. After more negotiation and delays, the necessary paper work was approved and both mother and son were allowed to leave during November, 1946. These events took place just before the Soviet Iron Curtain came down. The Russian wives of other Americans were not as fortunate. Those remaining were rounded up and sent to Siberia as "punishment" never to return to Moscow or their families. Uskievich knows of at least two Naval Officers whose wives were lost through this ruthless process.

While reading the above real-life experiences of the Uskievich family, Cmdr. Fredian recalled seeing an MGM movie "Never Let Me Go" in which a Russian ballerina married an American correspondent who was later deported. In the melodramatic movie all the exasperating, unsuccessful attempts to obtain an exit visa for his wife were detailed, and they finally resorted to an escape by sea. The 1953 movie starring Clark Gable and Gene Tierney was adapted from the novel *Two If By Sea* by Roger Bax (pseudonym) that was also issued in Great Britain as *Came The Dawn* in 1949 and later in 1986· under his real name Andrew Garve.

Sasebo.

Docking at Sasebo is recorded at 0910 on 6 January 1946. Care had been taken in bombing Sasebo not to obliterate the docks at the Japanese Naval Base, which is 34 miles NNW of Nagasaki. The fire bomb raid of 487 B-29 bombers on 28 June 1945, appeared to have had a devastating effect as a result of fire storms. I picked up a set of tea cups that fused together. Even tiles from the roofs showed signs of melting. During a brief walk away from the dock, I was confronted by a Japanese who bowed low and said "Welcome to Japan" in perfect English! That certainly dissipated any notions of being a conqueror! He explained that he had obtained a master's degree from Columbia University before the war. On learning of my interest in geology, he ran up a steep hill to his home and retrieved a handful of polished stones (Niki Ishi), which were presented as a gift.

There were other efforts by the Japanese to seek friendship with the Americans. One shop sold small "friendship" lapel pins that depicted Japanese and American flags side-by-side. For some in the unit it seemed premature to begin the healing process. Capt. Cumberledge was particularly irritated by the pin as he recalled all too vividly the loss of many friends.

In a surviving shop near the docks, there was a polished 8-inch sphere of a local rock. The method of grinding and polishing intrigued me and I wanted to buy it. Major Bristor reminded me that I had tried many persuasive arguments, but the shopkeeper would not part with it. A year or so later, an old high-school friend, then Lt. Richard L. Park, who was serving as military governor of one of the Prefectures, told me that those spheres (called suishō-tama) were family heirlooms and passed down from generation to generation. In due course, he managed to send me a 1.625-inch sphere ground from a natural, single, quartz crystal[68] showing a few

---

[68] The US National Museum of Natural History has a quartz sphere (crystal ball) on exhibit that was cut in China from a Burmese quartz crystal and ground and polished in Japan. It is flawless and measures 12.875 inches in diameter and weighs 106.75 pounds. An exquisite sphere was engraved in Russia during the last century from a single crystal of quartz from the Ural Mountains in the form of a statue of Atlas holding up the world. The piece is 4.625 inches high and is on display at the American Museum of Natural History in New York City.

fluid inclusions and lineage boundaries. One of the processes of grinding is quite simple but time consuming: merely rolling a roughed-out rock or crystal between two parallel steel plates with water and abrasive. It is done in much the same fashion as one rolls out a ball of dough in bread making. The ancestors represented by the quartz sphere, displayed in an appropriate setting, are respected today even though displaced from their family and native land.[69]

Near the docks of the Japanese Naval Base there was an underground bunker used as a command post. It was beautifully paneled in fine woods, but most of the furnishings had been removed. Among the trashed maps and documents was a white, plastic Japanese radiosonde and a stock of rice-paper parachutes.[70] Presumably, the intent was to retrieve the radiosondes after the balloon burst at the highest altitude reached. The radiosonde, measuring 4×4×6 inches, appeared to be of high quality and manufactured with precision. There is no doubt that the Japanese also placed high value on accurate knowledge of the weather for military operations (e.g., Pearl Harbor attack). The Soviets had captured the Japanese weather code when they invaded Manchuria, according to Major General Belyakov (CO Report No. 19, 3 October 1945), which would have been a valuable asset for ascertaining the weather over the targets and verifying the forecasts during the proposed invasion. An alternative source of weather information that the Germans used effectively along the coast of Labrador was the installation of automated remote weather stations (Douglas, 1996).

One of the trashed maps (Japanese No. 210, 1: 1,437,000) found in the bunker would have been of great value if available for the invasion. It contained a summary of the surveys by the Imperial Japanese Navy of soundings around the Ryukyu Islands, including Okinawa, and the southern portion of Kyushu where the landings were to take place. The existence of a large number of detailed charts

---

[69] It is still difficult for the author to reconcile the delicate and sensitive arts of such great beauty produced by the Japanese with the ugly and vicious war they precipitated.

[70] The radiosonde and parachute have been deposited in the National Museum of Naval Aviation, Pensacola, FL, where there was an aerological exhibit in May, 1995.

were noted on the map. Additions and corrections had been made up to 1930.

After the U.S. Marine Corps troops (91 enlisted men and 6 officers of the Signal Battalion of the Fifth Amphibious Group) were boarded, the ship sailed at 1:02 P.M. on 7 January 1946, arriving in San Diego exactly 14 days later (see Table 3). It was anticipated by the group that the ship would stop briefly in Hawaii, but the port was too busy to accommodate an unscheduled stopover. En route, many of the unit's reserve personnel were calculating their discharge points to see how much longer they had to serve on active duty. Others were contemplating their new duty stations. Still others would admire the U.S. Treasury paychecks they had accumulated as a nest egg for use after they were separated from service. But it was the good Navy chow, warm temperatures, and rest that was most appreciated. Everyone in the MOKO Expedition had escaped safely with the loss of only one appendix! One is fortunate enough to survive a war, but it is not often one has a lifetime of experiences jammed into six months.

POSTSCRIPT

THE KHABAROVSK WEATHER CENTRAL had become operational in time for the planned invasion of Japan, if needed, in spite of all the delays generated by the Soviets. There is no doubt that the Soviets never wanted weather centrals manned by Americans in their country. It should be noted that the only other U.S. military units to serve in the Soviet Union in the entire course of World War II were the 1,300 men of the Army Air Forces stationed in Poltava, Ukraine, for the shuttle bombing of Germany (Lucas, 1970). On the 4th of March 1946 they even terminated all exchanges of weather observations. On 20 March 1946, Representative John Tabor (NY) asked for an investigation by the House Foreign Affairs Committee because the Soviet Government had been allowed to extend for another three months the operation of its radio station in the War Department Building. He objected to the continued transmission of coded messages to Moscow and opposed the extension even though the State Department approved.

The calm but determined leadership of Captain A. A. Cumberledge can account for much of the success of the expedition tested at every turn by the exasperating and frustrating actions of the Soviet liaison officers and their superiors. He was called to Washington immediately on return to the U.S. to report on the mission. (Cmdr. Butow dealt with the reassignment of personnel and delivery of equipment.)

The unwillingness of the Soviets to reach a decision at the local level was not only the result of fear of the consequences of failure but also from their inherent suspicion of each other even within the family. It is difficult to judge what portion of the failure of the system should be assigned to the structure of the system itself and how much is to be attributed to the inherited culture of the Russians. Even the sadness of their folk music, as beautiful as it is, reflects the generations of subjugation and deprivation. No member of the unit could accept any of the precepts of the system. Many were quite open about their disgust for the centralized Soviet bureaucracy,[71]

---

[71] A young Russian colleague, who is applying now for entry into the U.S., thought the rising bureaucracy and centralization in the U.S. might be approaching that

127

and vented their feelings loudly, especially as they passed through the Custom House on their way out of the country.[72] The unit had indeed experienced the "Cold War" in both temperature and political rivalry before the term was defined.[73]

On return to the U.S., however, it was not appropriate to even mention the fact that one had been in the USSR. For example, Senator Joseph R. McCarthy launched in 1950 an anti-Communist witch-hunt by making unsubstantiated accusations. He had little regard for civil liberties and disregarded, through sensationalism, the common rules of decency and truth. American society had generated its own aura of fear through anti-communism supported by a rash of anti-subversion laws and McCarthy-like state inquisitions. It is no wonder that the record of the MOKO Expedition was quickly submerged in the atmosphere created by McCarthyism. When the remainder of the thirty-nine reports of the CO are uncovered in the Archives, perhaps a more accurate account of the contribution the MOKO Expedition made toward the planned invasion of Japan can be evaluated. There is even some prospect that documents on the MOKO Expedition in the Russian files may become available for review.

Knowledge of the Siberian weather was essential to the success of the invasion, but one cannot even begin to contemplate the human cost of such a venture that was mercifully terminated by the use of atomic weapons.

---

in the old Soviet system—but he did have a smile on his face as he struggled with the paperwork.

[72] The record of those vociferous condemnations of the Soviet system appears to have been lost in the Soviet bureaucracy. The author was a guest of the Soviet Union in 1974 as a lecturer on experimental petrology at the universities in Moscow, Leningrad, Kiev, and Novosibirsk. No mention was made of the author's tour of duty in Siberia. One lecture, however, emphasizing the freedom researchers have in selecting and developing their own scientific programs in the U.S. (Yoder, 1990) was ordered not to be repeated in the universities yet to be visited. In addition, on 26 January 1977, the author was honored at a reception in the Embassy of the USSR in Washington and presented a diploma attesting to his election as an Honorary Member of the All-Union Mineralogical Society of the USSR. Apparently scientific knowledge was considered more valuable than political views to the Soviets some thirty years later. With the passage of fifty years, Russian scientists are being warmly welcomed in U.S. institutions without regard for their political views.

[73] The term "Cold War" was first used during the congressional debates of 1947 by presidential advisor Bernard Baruch. The confrontation lasted until 1989!

ACKNOWLEDGMENTS

MANY OFFICES AND AGENCIES contributed to the documentation of the events described: Naval Historical Center (Washington, DC); Public Affairs Office, Naval Meteorology and Oceanography Command, Stennis Space Center, MS; Fleet Numerical Meteorology and Oceanography Detachment, Asheville, NC; Hoover Institution; Chief Historian Joint Chiefs of Staff; Historian and Librarian, State Department ; Air Weather Service, Scott Air Force Base, IL; National Weather Service; National Institute of Science and Technology; National Climatic Data Center, U. S. Information Agency; Historian, McDonnell-Douglas Aircraft Co.; Naval Weather Service Association; National Ocean Service and the Library of the National Oceanic and Atmospheric Administration; and the Center for Air Force History, Bolling AFB, DC.

The first draft of the manuscript was entered on the word processor by Mrs. Susan A. Schmidt in addition to her regular duties. Her kindness and patience in making the multitude of inserts and corrections as recall was stimulated by the reviewing process was most appreciated.

Critical reviews were volunteered at an early stage by Drs. Robert M. Hazen and Russell J. Hemley and Mr. Shaun Hardy of the Geophysical Laboratory and Mrs. Daniela Power of the Department of Terrestrial Magnetism. Additional detailed reviews at various later stages were made by Mr. Jack Almquist, Dr. Louis Brown, Dr. Nabil Boctor, Prof. Paul Bushkovitch, Cmdr. Thomas V. Fredian, Dr. Alexander Goncharov, Mr. Jack A. Green, Capt. George J. Haltiner, Dr. J. Marvin Herndon, Prof. Dr. H. G. Huckenholz, Miss Marjory Imlay, Prof. Martin L. Levitt, Dr. Bjorn Mysen, Dr. Charles T. Prewitt, Mr. Raymond A. Rzeszut, Mr. William L. Smallwood, Mr. John Straub, Mrs. Merri Wolf, and Dr. Ellis L. Yochelson. Their wide range of perceptions was of considerable help in preparing the final draft. It is a pleasure to acknowledge their efforts in clarifying the description of events in this little known area of the world.

A current colleague at the Geophysical Laboratory, Dr. Ho-kwang Mao, provided first-hand insights into the Japa-

nese occupation of China. His father, Major General Mao Sen
(written Chinese style) was very active in the guerrilla forces
in the region of Shanghai and awarded the Legion of Merit
by the President of the United States.

A penultimate draft was reviewed by a Japanese friend
born just before the war, Prof. Ikuo Kushiro, who has been a
Visiting Investigator at the Geophysical Laboratory and subsequent winner of the Japan Prize. He noted that there was
little or no residual resentment of Americans who are more
often remembered by the Japanese people for their help in
resurrecting Japan as a nation after the war.

Another Japanese friend of long standing, Prof. Kenzo
Yagi, commented on the manuscript, and responded on the
51st anniversary of the unconditional surrender of Japan
"with deep feeling."[74]

Finally, it is a pleasure to acknowledge the enthusiastic
help of our Librarian Shaun Hardy and his assistant Mrs.
Merri Wolf. Their broad knowledge and ingenuity were critical to acquiring quickly the wide range of reference materials
needed for documentation.

---

[74] He wrote recently, "... at noon on 15 August, we heard by radio the Emperor's
edict that the terms of unconditional surrender were accepted. Deep sighs came
out of the group. Now the war was finally ended by the defeat of Japan. We had
complex feelings, a mixture of grief and relief." Acceptance of the decision was
apparently immediate as he rode the train to his home, which survived the fire-
bomb raid on Sendai: ... "we could see the lights in every house, since light
control was abolished. These lights, though not so bright, gave us the strong impression that the war is now ended, and we came back to peace!" In 1949, Dr.
Yagi was one of the first group of 30 young professors from occupied Japan who
were sponsored by the U.S. Army and the Institute of International Education
for study of the modern methods used in the U.S. After attending the Colorado
School of Mines for a year he became a Visiting Investigator at the Geophysical
Laboratory during 1950–51. After an initial, but brief, awkward period on the
arrival of the "enemy," his outgoing personality and incredible skills not only in
science, but also in the arts and social amenities, won many friends for the new
Japan. He published joint experimental investigations with staff members, including the author. He quickly adopted many of the American mannerisms, in
fact, so many that he had to be cautioned, when he returned to Japan, to abide
by the cultural mores of his elders. Needless to say, we became close personal
friends and visited each other's homes in succeeding years as well as shared our
common interest in volcanoes.

APPENDIX A
BIOGRAPHY OF COMMANDING OFFICER FLEET
WEATHER CENTRAL, KHABAROVSK, USSR

CAPTAIN ARTHUR ALBERT CUMBERLEDGE was born in New Castle, PA, on 24 May 1908, son of Albert O. and Olive M. Cumberledge. After residing in Farrell, PA, his parents moved to Youngstown, OH, when he was eight years old. He entered the U.S. Naval Academy in 1927 and graduated with honors in 1931. He served on destroyers, battleships, and aircraft carriers before going to the Post Graduate School in Annapolis in 1938 for a three-year course on General Line and Aerology. The third year was spent at the Massachusetts Institute of Technology where he received an M.S. degree in meteorology in 1941. Thereafter he joined the USS Ranger (CV4) and was then transferred to the USS Hornet (CV8) on which he prepared the weather forecast for the Doolittle bombing raid (16 B-25s) on Tokyo on 18 April 1942. The USS Hornet was sunk on 27 October 1942 near the Santa Cruz Islands. He managed to swim for hours, a skill developed at the Naval Academy as an outstanding water polo player, before being rescued by a U.S. destroyer. After service on the staff of Adm. Halsey in Noumèa, New Caledonia, Cumberledge returned to the U.S. and was in the Office of the Chief of Naval Operations from 1943 to 1945, serving as Assistant Director and Detail Officer in the Aerological Section. In July 1945, he was spot promoted to Captain and ordered to be the Commanding Officer of the Advanced Base Unit MOKO. For his outstanding service in the MOKO Unit, later called the Fleet Weather Central Khabarovsk, he received the Bronze Star. Subsequent duties involved assignments at the Fleet Weather Centrals in San Diego, CA, Pearl Harbor, HI, and Yokosuka, Japan. In addition, Capt. Cumberledge served as Head of the Aerological Engineering Department at the U.S. Naval Post Graduate School in Monterey, CA. He was on the staff of Adm. Blandy for Operations "Crossroads" for the Bikini atom bomb tests and received a Commendation Ribbon from the Secretary of the Navy for his contributions. After a two-year tour at the Fleet Weather

131

Central in Guam, he was transferred to London, England, to serve as Aerological Staff Officer for the Commander in Chief of Naval Forces in Eastern Atlantic and the Mediterranean. He retired from the U.S. Navy on 1 July 1960.

Captain Cumberledge died of cancer and heart disease on 1 January 1996 at the age of 87. His ashes were scattered at sea.

APPENDIX B
SUMMARY OF WEATHER OBSERVATIONS

THE WEATHER IN THE KHABAROVSK AREA was of special interest to Major C. L. Bristor, and he prepared and preserved a nine-page report[75] on the winds, fronts, storm tracks, precipitation, and flying conditions. On the basis of the hourly observations, twelve-hour pilot balloon and radiosonde soundings at the Fleet Weather Central as well as the data provided by the Red Army Meteorological Office in Khabarovsk, he personally made a re-analysis of the official six-hourly weather maps. For example, he made tracings of the weather maps of a selected area pertinent to the mission, and made alterations to improve the continuity. Fortunately, he was able to smuggle out of the USSR some ninety tracings of weather maps, one for each day of the period 15 September to 15 December 1945. In addition, he had constructed some time cross-sections that were most valuable in marking frontal passages. Nevertheless, he was reluctant to draw any firm conclusions because of the limited amount of data and the short time period.

Surface and gradient winds.

Major Bristor investigated the relationship between the pilot balloon winds at gradient level and the surface winds to see if the surface winds were representative. He noted that some 158 pilot balloon soundings had been made. The hydrogen gas for the balloons was generated with a device developed by Pan American Airways Inc., which had operated its own weather network in the Caribbean Sea area. The gradient winds from the south-southwest through west were favored (Fig. 50). Apparently there was a semi-permanent trough along the Amur River during that season. The surface winds estimated visually from a microanalysis of the isobaric pat-

---

[75] The unpublished report provided by Major C. L. Bristor is entitled "Synoptic processes in Central Siberia during the fall season 1945." The ninety tracings of weather maps are in a folder marked "Surface weather charts—analysis by C. Bristor at FWC Khabarovsk."

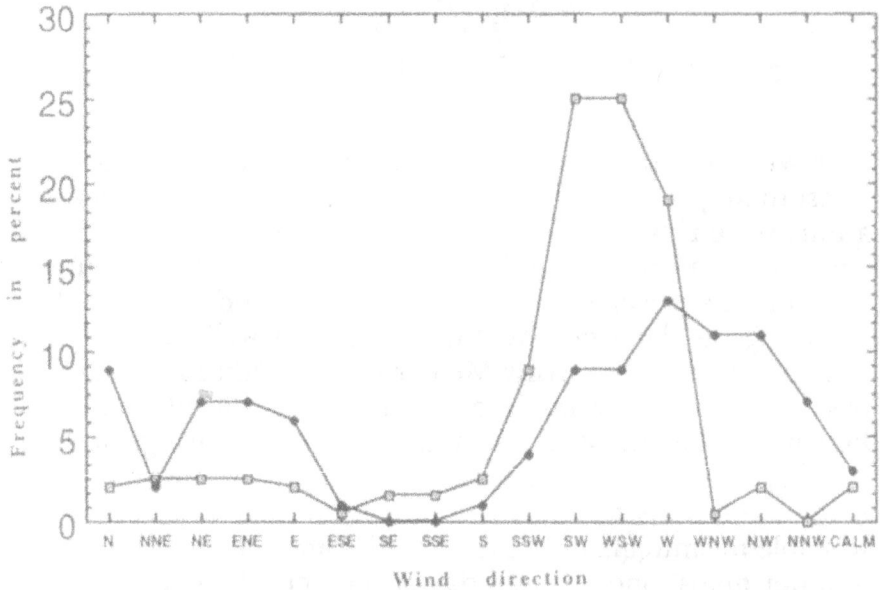

Fig. 50. Frequency of wind directions observed during the period 15 September to 15 December 1945 at the Fleet Weather Central Khabarovsk. Open squares from 158 pilot balloon soundings. Solid diamonds from 90 tracings modified from official 24-hourly weather maps. Compiled by Major C. L. Bristor.

Fig. 51. Microanalysis by Major C. L. Bristor of the isobars at one millibar intervals in the region of the lower end of the Amur River basin and the Sikhota Alin mountains. A typical pattern resulting from an intense low-pressure cell in the Sea of Okhotsk and a strong gradient of pressure increasing to the west.

tern, with due regard for the frictional coefficient of drag, tended to reflect the gradient winds quite closely. Bristor was of the opinion that the pattern exhibited in Fig. 51 would exist on every weather map where lower pressure exists in the Sea of Okhotsk. He cautioned, however, that the temperature factor in reduction of pressure may be extreme and lead to large errors. That is, the ridge of pressure parallel to the coast may result in part from the pressure reductions at stations in the higher elevations of the Sikhota Alin mountains. These observations reinforce the view that detailed knowledge of the specific conditions at every station is required to produce a meaningful weather map.

Troughs and fronts.

The frequency and duration of weather associated with troughs and fronts is often the critical issue in military operations. In the three-month period, an average of 14 systems passed over Khabarovsk each month, but only a third of these were sharply defined. The troughs aloft, presumably persisting from old fronts, passed over the Siberian high-pressure air mass, rarely penetrating to the surface. According to Bristor, 42 percent of the hours observed could be classed as cloudy; these periods lasted most frequently from 10 to 20 hours. Two periods of five and seven days of clear or scattered clouds resulted from prolonged northeasterly winds that were generated by two intensive lows in the Sea of Japan from typhoons moderating to extra-tropical storms.

Storm tracks.

Bristor plotted the storm tracks of some thirty centers that moved eastward off the Siberian landmass (Fig. 52). About a dozen of these low-pressure centers originated in the region of Lake Baikal, usually as a spinoff of older systems. Four came out of Manchuria or Eastern Mongolia. The remaining lows developed as waves on southwestward tracking cold fronts. In general, the path of the storm tracks curved northeastwardly to the Sea of Okhotsk whereas a dozen trended easterly or southeasterly. The fronts trailing from the center of the low-pressure centers would have been the dominant

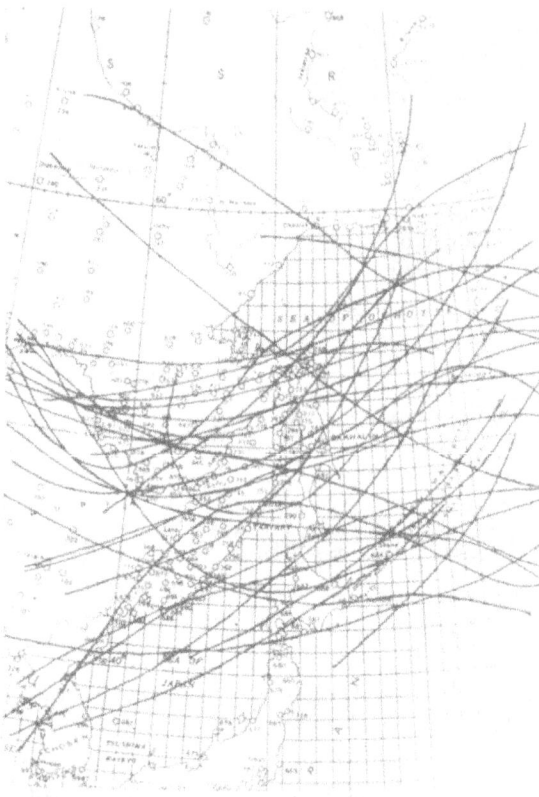

Fig. 52. Storm tracks, recorded by Major C. L. Bristor during the period 15 September to 15 December 1945, generated on the Siberian land mass, in Manchuria and Eastern Mongolia, and from secondary waves on trailing fronts.

weather makers during the invasion because of the strong contrast of air masses marked by the frontal boundary.

Temperature.

The record of daily observations taken at the MOKO weather station were apparently destroyed before leaving. None of the officers located retained copies of the official records, and the data received at Guam were not retrievable if archived. Some measure of the range of temperatures can be gained from the climatic summary at Khabarovsk (Anonymous, 1966) for the period 1914–1921 (Fig. 53). The data for the period 1951–1994 in the World Climate Data Bank show closely similar trends; however, the absolute maximum and minimum were 104°F and −45°F, respectively, compared with 91°F and −46°F for the earlier seven-year period. The continuous and substantial drop in temperature during the

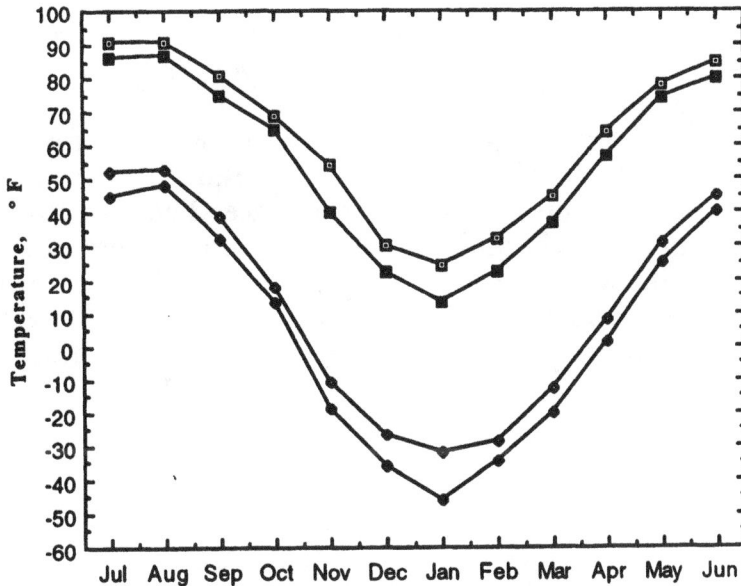

Fig. 53. Temperatures recorded at Khabarovsk during 1914–1921. Upper curve with open squares, absolute maximum. Closed squares, average of highest temperatures reached each month. Open diamonds, average lowest temperature reached each month. Closed diamonds, absolute minimum.

period the Fleet Weather Central Khabarovsk was operational is evident.

Precipitation.

There was record of precipitation during less than 9 percent of the total hours of record. Slightly over 4 inches of precipitation occurred during the three-month period with the bulk of it falling in September as rain. (Note that 12 inches of freshly fallen dry snow is equal to about one inch of rain.) Snow flurries lasted less than four hours and there were only four cases of 10 or more hours duration. The longest snow period was 24 hours and was associated with a strong low in the Sea of Okhotsk. Surprisingly enough in that case, the snow fell from clouds that were well above 5,000 feet! In one twelve-hour period during the middle of October, fog and drizzle brought the ceilings down to 500 feet as a front stalled over the city. Icing was considered to be a negligible problem except in a few isolated frontal situations. It is noted that during mid October, frost was an important problem for planes left out of the hangar.

The several lengthy rainy periods in September generated very muddy and slippery roads, presumably because of the high clay content of the road surface. The snow covers were generally light and dry so that the roads were cleared with a light wind. Only three cases were noted where the snow accumulated to two or three inches. Blowing snow induced one to wear a face mask because of the ice needles.

Flying weather.

It was concluded that the flying weather in the Khabarovsk area was excellent most of the time period. For 34 clear periods, the average duration was 33 hours, more than adequate time for shuttle or return bomb runs to Japan. Even on cloudy days the clouds were in the middle or high layers. Low clouds occurred only 19 percent of the hours in the three-month period. Ceilings below 1500 feet were less than 2 percent of the entire record.

(Major Bristor was such a dedicated and inveterate weatherman that he kept a record of the weather after the Weather Central was closed, during the trip to Vladivostok, and each day at sea en route to San Diego. Because of his "unremitting industry and analytical experience" he was awarded a Commendation Ribbon by the Commander-in-Chief of the Pacific Fleet. Fortunately for the nation, he returned to work at the U.S. Weather Bureau after discharge from the army and an additional year at MIT.)

REFERENCES

Adamson, H. C. and G. F. Kosco (1967). *Halsey's typhoons: A firsthand account.* New York: Crown Publishers, Inc., 206 pp.

Anonymous (1871). "The practical use of meteorological reports and weather-maps." Office of the Chief Signal Officer. Circular. Washington: Government Printing Office, 37 pp.

Anonymous (1943) "Meteorological data for Japan." H. O. Misc. No. 10,638. Washington: Hydrographic Office, 130 pp.

Anonymous (1944) "Breakers and surf: Principles in forecasting." H.O. No. 234. Washington: Navy Department, 61 pp.

Anonymous (1945). "The USSR Institutions & People: A brief handbook for the use of officers of the armed forces of the United States." Washington: Government Printing Office, 130 pp.

Anonymous (1955) "The entry of the Soviet Union into the war against Japan: Military plans, 1941–45." Washington, DC: Department of Defense, 107 pp.

Anonymous (1957) "Winds, weather and warships." Naval Aviation News, October No., pp. 1–7.

Anonymous (1966) "Tables of temperature, relative humidity, and precipitation for the world." Part V, Asia. Meteorological Office. London: H.M. Stationary Office. p.18.

Anonymous (1989) "Honour in the Jungle," *The Economist,* 313, No. 7624, (October 14, 1989): 40.

Armstrong, A. (1961) *Unconditional surrender: The impact of the Casablanca policy upon World War II.* New Brunswick, N. J.: Rutgers University Press, 304 pp.

Bates, C. C. and J. F. Fuller (1986) *America's Weather Warriors, 1814–1985.* College Station: Texas A & M University Press, 360 pp.

Bax, Roger (1949) *Two If by sea.* New York: Harpers & Brothers, 245 pp.

Brandes, H. W. (1826) "Dissertatio physica de repentinis variationibus in pressione atmosphaerica observatis." Leipzig.

Brown, F. A. (1962) "U. S. Navy Weather Stations in Siberia." *Proceedings U.S. Naval Institute,* 88, No. 7: 76–83.

Byers, H. R. (1970) "Recollections of the war years." *Bulletin American Meteorological Society,* 51. No. 3: 214–217.

Byers, H. R. (1976) "The founding of the Institute of Meteorology at the University of Chicago." *Bulletin American Meteorological Society,* 57, No. 11: 1343–1345.

Carter, K. C. and R. Mueller (1973) "The Army Air Forces in World War II: Combat Chronology 1941–1945." Washington:

Albert F. Simpson Historical Research Center, Army Air Forces, 991 pp.

Chuiko, V. V. "Complete works of M. Y. Lermontov." 2 vol. in 1. St. Petersburg: M. O. Wolf Association: 303; 366 pp.

Cohen, S. B. (1981) *The forgotten war: A pictorial history of World War II in Alaska and Northwestern Canada.* v 2. Missoula, MT: Pictorial Histories Publishing Co. 254 pp.

Craven, W. F. and J. L. Cate (1983) The Army Air Forces in World War II" vol.5 *The Pacific: Matterhorn to Nagasaki June 1944 to August 1945.* Chicago: University of Chicago Press, 878 pp.

Crowell, J. C. (1946) "Sea, swell, and surf forecasting methods employed for the Allied invasion of Normandy, June 1944." M. A. Thesis. Los Angeles: University of California, 81 pp.

Cumberledge, A. A. (1946) "Navy establishes fleet weather centrals in Siberia." Remarks recorded at press conference on 3 March 1946. On file at Naval Meteorology and Oceanography Command, Stennis Space Center, Bay St. Louis, MS. 6 pp.

Deane, J. R. (1947) *The strange alliance: The story of our efforts at wartime co-operation with Russia.* New York: Viking Press, 344 pp.

Douglas, W. A. B. (1996) "Beachhead Labrador." MHQ: *Quarterly Journal of Military History,* 8, No. 2: 35–37.

Ferrians, Jr., O. J., R. Kachadoorian, and G. W. Greene (1969) "Permafrost and related engineering problems in Alaska." Geological Survey Professional Paper No. 678. Washington: Government Printing Office.

Fowle, B. W. (1992) "Builders and fighters: U.S. Army Engineer in World War II." Fort Belvoir, VA: Office of History, U.S. Army Corps of Engineers, 529 pp.

Glantz, D. M. (1983) "August storm: The Soviet 1945 strategic offensive in Manchuria." Leavenworth Papers No. 7. Leavenworth, KS: Combat Studies Institute, 205 pp.

Graves, W. S. (1931) *America's Siberian Adventure, 1918–1920.* New York: J. Cape & H. Smith, 231 pp.

Hansell, H. S., Jr. (1980) "Strategic air war against Japan." Washington, DC: Government Printing Office, 72 pp.

Harriman, W. A. and E. Abel (1975) *Special Envoy to Churchill and Stalin, 1941–1946* New York: Random House, 596 pp.

Lauderbach, R. E. (1946) *Through Russia's Back Door.* New York: Harper & Brothers Publishers, 239 pp.

Lucas, R. C. (1970) *Eagles east: The Army Air Forces and the Soviet Union, 1941–1945.* Tallahassee: Florida State University Press, 256 pp.

March, G. P. (1988) "Yanks in Siberia: U.S. Navy weather stations in Soviet East Asia 1945." *Pacific Historical Review*. Berkeley,CA: University of California Press, pp. 327–342.

Mendeleyev, D. I. (1865) "Ueber die Verbindung des Weingeistes mit Wasser." *Zeitschrift für Chemie* I: 257–264. (Translation in part of doctor's dissertation, St. Petersburg. A complete translation from the original Russian is given in *Annalen der Physik und Chemie*, 138 [1869]: 103–141, 230–279.)

Miles, M. E. (1967) *A different kind of war: The little-known story of the combined guerrilla forces created in China by the U.S. Navy and the Chinese during World War* II. Garden City, NY: Doubleday, 629 pp.

Molnar, G. W. (1946) "Survival of hypothermia by men immersed in the ocean." *Journal of the American Medical Association*, 131: 1046–1050.

Myles, B. (1981) *Night Witches: The untold story of Soviet women in combat*. Novato, CA: Presidio Press, 278 pp.

Shalett, S. (1946) "Soviet weather contact cut; U.S. Siberia stations closed." New York: *The New York Times*, 21 March 1946, pp. 1, 6.

Skates, J. R. (1994) *The invasion of Japan: Alternative to the bomb*. Columbia, SC: University of South Carolina Press, 276 pp.

Stratton, R. O. (1950) *SACO, the rice paddy navy*. Pleasantville, NY: C. S. Palmer Publishing Co., 408 pp.

Sverdrup, H. U. and W. H. Munk (1943). "Wind, waves and swell: A basic theory for forecasting." Wave Report No. 1. La Jolla, CA: Scripps Institution of Oceanography, 130 pp. (Revised in 1947 as US Hydrographic Office Technical Report No. 1, 44 pp.)

Swanborough, G. and P. M. Bowers (1968) "United States Navy Aircraft Since 1911." London: Putnam, pp. 169–170.

Uznanski, M. E. (1957) "The stamp '*Dojdziemy*': The Polish Field stamp in Russia." *Bulletin Polonus Philatelic Society*, No. 150, pp.1–16.

Voronov, N. N. (1965) "Exploits of the Soviet Nation." *Istoriia SSSR*, No. 4, pp. 13–27 (in Russian).

Weintraub, S. (1995) "The three-week war." *MHQ: Quarterly Journal of Military History*, 7, No. 3: 86–95.

Wilkins, M. and F. E. Hill (1964) *American Business Abroad: Ford on Six Continents*. Detroit: Wayne State University Press, 541 pp.

Williams, M. H. (1960) "Chronology 1941–1945" in *United States Army in World War II, Special Studies*. Washington, DC: Department of the Army, 660 pp.

Willis, C. (1988) "History of the U.S. Fleet Weather Centrals USSR" *Naval Oceanography Command News*, 8, Nos. 6–7: 12–15.

Winick, L. (1996) "Japan's 1945 'Enemy Country, Surrender' Stamp." *Scott Stamp Monthly*. September, p. 56.

Yoder, H. S., Jr. (1990) "The CIW: A model of scientific freedom." *The World & I.*v. 5, No. 1: 384–393.

Zawodny, J. K. (1962) *Death in the forest: The story of the Katyn Forest massacre.* Notre Dame, IN: University of Notre Dame Press, 235 pp.

TABLE 1

List of Personnel

US Naval Advanced Base Unit MOKO

OFFICERS

Commanding Officer:	Capt. A. A. Cumberledge
Executive Officer	Cmdr. E. E. Butow
Communications Officer	Cmdr. A. K. Kinch
Supply Officer	Lt. Cmdr. E. K. Outcalt
Aerologists	Lt. P. P. Starke
	Lt. H. S. Yoder, Jr.
	Lt. (jg) R. W. Monical
	Lt. (jg) F. A. Nelson
Medical Officer	Lt. (MC) J. W. Revere
Liaison Officer	Lt. P. Worchel*
Disbursing Officer	Lt. (jg) B. Uskievich*
Communications	Lt. J. T. Shaw
	Lt. J. S. Rhodes*
	Ens. R. L. Hubbard*
	Ens. L. W. Bowden*
	Ens. C. Birkett
US Army Liaison	Major J. W. Kodis
	Major C. L. Bristor

NON-COMMISSIONED OFFICER

Ship's Clerk	Warrant J. Klopovic*

ENLISTED MEN

Chief Aerographer's Mate	E. F. Lewis
	G. F. Wall
Aerographer's Mate 1c	H. A. Barger
	G. P. O'Donnell
	H. A. Slauenwhite, Jr.
	S. E. Wyers
Aerographer's Mate 2c	C. W. Buxton
	R. C. Gotthardt
	S. B. Lee
Aerographer's Mate 3c	G. Beniash
	J. C. Gillman, Jr.
	J. B. Hodges
	A. J. Medina
	W. G. Morgan

143

Table 1 (continued)

	G. W. Pope
	J. G. J. Thomas
Pharmacist's Mate 1c	C. O. Adams
Radiomen 1c	F. H. Hudson
	A. R. Locatelli
Radiomen 2c '	W. F. McNeely
Radiomen 3c	G. A. Breiten
	A. V. Smith
Radio Technician 3c	D. W. Conard
	M. H. Doty
	C. G. Flittie
	N. E. Pembrock
	S. Resnick
	B. M. Schmeltzer
Communication Yoeman	A. G. Sparks
Motor Machinist's Mate 1c	E. W. Chesney
Motor Machinist's Mate 3c	W. E. Beasley
Specialist (X) 3c	G. M. Golovin
Storekeeper, disbursing 1c	E. A. Hazzard
Yoeman 1c	J. H. Louk
Yoeman 2c	W. H. Gibson
Seaman 1c	R. K. Fisher
	S. J. Maddox
	H. B. Phillips
	J. P. Sullivan
	R. J. Werhle
Seaman 2c	J. E. Cannon

c = Class
* = Russian speaking

TABLE 2

List of Soviet Civilian Personnel

Officer's Galley
  Cooks: Gotovchenko*
                                            Kulagina
  Waitresses: Surikova, Nadya
                                            Sholudyakova, M.*
  Scullery Maids: Nogunova
                                          Gavlenko

Men's Galley
  Cooks: Hoolosova
                                          Kleshchevneikoa
  Waitresses: Nosikova
  Scullery Maids: Goodkova
                                          Sorokina
  Housekeeper: Bragina

Maids at Buildings
  Dachas: Zorina, Maria Alexeevna*
  Administrative Building: Orefyeva, Klava
  Men's Barracks: Fomina, Antonina

Services
  Laundry: Prikhodko
  Barberess: Monitz
  Stokers: Tatyanushkn
                                          Leonov

*Commended by Capt. Cumberledge on departure of the Unit

TABLE 3

Timetable of Events

US Weather Programs in Siberia

1941		
	Sept	Weather exchange program initiated with the USSR
	7 Dec	Japanese attack on Pearl Harbor, HI
1943		
	Feb	Weather exchange program with Soviets expanded
1944		
	17 Oct	Stalin agrees to Manchurian offensive three months after the defeat of Germany
	18 Dec	Typhoon Cobra severly damages Halseys' Fleet
1945		
	Feb	Offer made by U.S. at Yalta Conference to place weather equipment and personnel in Siberia
	3 Apr	Plans for invasion of Japan formulated
	5 Jun	Typhoon Viper damages U.S. Third Fleet
	18 Jun	Invasion plans approved by President Truman
	Jul	Permission granted by USSR at Potsdam Conference to place two Navy weather stations in Siberia
	2 Aug	U.S.-manned weather stations in Siberia approved at Potsdam Conference
	4 Aug	Orders cut for U.S. personnel to man two Siberian weather stations
	6 Aug	Atomic bomb dropped on Hiroshima
	9 Aug	USSR declares war on Japan and invades Manchuria
		Atomic bomb dropped on Nagasaki
	14 Aug	Japanese accept terms of "unconditional surrender"
	15 Aug	All U.S. offensive action against Japan is terminated
	17 Aug	USSR reaffirm arrangements for Siberian weather stations

TABLE 3 (continued)   147

18 Aug	Personnel assembled in Seattle informed of mission to Siberia
20 Aug	Lend-lease shipments to USSR terminated
22 Aug	Soviets abandon plans to invade Hokkaido from Sakhalin
23 Aug	First flight of Moko Unit takes off from Fairbanks, AK for Siberia
26 Aug	First flight of Moko Unit arrives in Khabarovsk, Siberia
28 Aug	Meetings with Soviet liaison and weather officers at MOKO base
	Advance party of U.S. troops land on Honshu Island
30 Aug	U.S. troops land in force on Honshu Island
6 Sept	Third flight of Moko Unit arrives in Khabarovsk
	Personnel landed at Petropavlovsk by ship to set up weather station
18 Sept	Soviets demand copies of all ciphers and codes used
22 Sept	Name changed from "Advanced Base Unit MOKO" to "Fleet Weather Central Khabarovsk"
26 Sept	Radio teletype placed in operation with Guam
27 Sept	Admiral King rejects Soviet demand for ciphers and codes
	Liberty for all U.S. personnel restricted by Soviets
28 Sept	Fourth flight of Moko Unit arrives in Khabarovsk
9 Oct	Typhoon passes over Okinawa, prospective staging area for planned invasion
12 Oct	Responses regarding ciphers and codes unresolved
14 Oct	Radio communications established with weather station at Petropavlovsk
15 Oct	First weather bulletins transmitted to Guam
17 Oct	U.S. personnel on liberty confronted by local and military police officers
25 Oct	Identification papers issued to all U.S. Personnel
28 Oct	U.S. Navy Day party attended by Soviet military officers
1 Nov	Date planned for initial phase of invasion of Japan
22 Nov	Soviets photograph MOKO installations and personnel
5 Dec	General Ivanov "requests" that U.S. Weather stations be closed
11 Dec	Orders received from Chief of Naval Operations to close down weather units by 15 Dec 45

Table 3 (continued)

15 Dec	Operations at U.S. Weather Central Khabarovsk terminated as ordered
19 Dec	Petropavlovsk weather station evacuated by USS Sarsi
21 Dec	Mayor of Khabarovsk visits MOKO base at Knaz Volkonka
28 Dec	Personnel and equipment depart Khabarovsk by train to Vladivostok
30 Dec	Permission granted for U.S. personnel to board USS Starr in Vladivostok after customs inspection
31 Dec	Hostilities with Japan terminated by Presidential Proclamation
1946	
2 Jan	USS Starr departs from Vladivostok for Sasebo, Japan
7 Jan	USS Starr departs Sasebo, Japan, for San Diego, CA
21 Jan	MOKO Unit arrives in San Diego, CA
1951	
8 Sept	Military occupation of Japan formally ended by Treaty of San Francisco (Final ratification 28 April 1952)

INDEX

149

www.ingramcontent.com/pod-product-compliance
Lightning Source LLC
Chambersburg PA
CBHW080926100426
42812CB00007B/2381